Relocating England

Published in our
centenary year
2004
MANCHESTER
UNIVERSITY
PRESS

To the memory of my father

Relocating England

P. W. Preston

Manchester University Press
Manchester and New York

distributed exclusively in the USA by Palgrave

Copyright © P. W. Preston 2004

The right of P. W. Preston to be identified as the author
of this work has been asserted by him in accordance
with the Copyright, Designs and Patents Act 1988.

Published by
Manchester University Press
Oxford Road, Manchester M13 9NR, UK
and Room 400, 175 Fifth Avenue, New York, NY 10010, USA
www.manchesteruniversitypress.co.uk

Distributed exclusively in the USA by
Palgrave, 175 Fifth Avenue, New York,
NY 10010, USA

Distributed exclusively in Canada by
UBC Press, University of British Columbia, 2029 West Mall
Vancouver, BC, Canada V6T 1Z2

British Library Cataloguing-in-Publication Data
A catalogue record for this book is available from the British Library

Library of Congress Cataloging-in-Publication Data applied for

ISBN 0 7190 6934 3 *hardback*
EAN 978 0 7190 6934 5
ISBN 0 7190 6935 1 *paperback*
EAN 978 0 7190 6935 2

First published 2004

13 12 11 10 09 08 07 06 05 04 10 9 8 7 6 5 4 3 2 1

Typeset by Helen Skelton, Brighton, UK
Printed in Great Britain
by Biddles Limited, King's Lynn

Contents

	Preface	*page* vi
	Acknowledgements	vii
1	Relocating England	1
2	Political-cultural identity	11
3	A development history	38
4	The project of Britain	49
5	British political discourse	81
6	Shock and price	106
7	First implications	116
8	The European Union project	130
9	Storytelling I: discourses of England	157
10	Storytelling II: discourses of Europe	179
11	Afterword: subaltern dreams	204
	Bibliography	209
	Index	219

Preface

Prien am Chiemsee is a small town in southern Germany, about eighty kilometres south of Munich. It is a pretty town. It has a mix of small farms located right in the town itself, a compact core with a church, shops and cafes around a small square, a scatter of elegant houses leading down to the lake and a wider spread of newer housing making up the suburbs. It is mostly a resort town now. It is also home to a Goethe Institute. I was a student there, struggling with *Mittelstufe* in the company of eight or nine participants, Italian, Greek and Icelandic. The class was small, our teacher marvellous. It was late in the year, there was light snow on the ground, and after classes it was pleasant to walk, to go down to the lakeside cafes, to watch the old-fashioned steam launches plying their last tourist runs and to eat fresh trout with horseradish sauce. It felt like the end of a season. It was. It was November 1989 and in the company of my elderly landlady I watched the events of that autumn unfold. She was astounded and delighted and her excitement made her German too quick for me to follow easily. But I was pleased for her. I was also pleased for me as it is not often one has the chance to watch the wheel of history turn. And turn it did; undermining, thereby, a spread of received ideas. This little book traces some of the still unfolding consequences of the events that I watched from that snowy, pretty town.

Acknowledgements

One's home country looks different seen from outside the bubble of its own self-understandings. And this text has been written in the main whilst travelling. Some work was done in the early 1990s and I have debts to colleagues and students in universities in Germany, Japan, Australia, Singapore and Malaysia. I have acknowledged them directly in other texts; once again my thanks. The final text has been assembled recently. The University of Birmingham kindly granted me leave in 2002–03, which enabled me to take up a number of guest professorships (one might call it a 'busman's sabbatical'). I spent the autumn at the Hamburg University for Economics and Politics, located in a beautiful old part of the city near the Alster Lake, where I worked with Professor Dr Wolfgang Voegeli. As ever, being in Germany was a pleasure and I offer my thanks to my colleagues and students. I should like to thank Elisabeth Niesing for the motto/joke that serves to begin this text. Over Christmas and New Year I spent time at the Institute for International Studies at Ramkhamhaeng University, located in the busy heart of Bangkok, and I am happy to record my thanks to the director, Dr Piboon, and my students. I spent the spring semester in the Department of Government and Public Administration of the Chinese University of Hong Kong, located in the quiet hills overlooking Tolo Harbour (thus escaping the UK in the early phase of the Iraq war, only to find myself in a community suffering from an epidemic of an hitherto unknown disease – something which recalled childhood in Yorkshire (outbreaks of smallpox, quiet panic and queuing for inoculations (happily not necessary with SARS)). I should like to thank Professor Kuan, his colleagues and my students for their welcome and for making my somewhat distracted visit happy and productive.

Motto / joke

There are three restaurants; French, German and British. A man decides to take lunch.

He goes into the first restaurant, orders food, the waiter brings it and says 'Bon Appetit!' He goes into the second, orders food, the waiter brings it and says 'Gut Essen!' And finally he goes into the third, orders food, the waiter brings it, looks at the man, looks at the food, and says 'Never Mind!'

1 Relocating England

At the time of writing – the early years of the twenty-first century – the structures of power[1] that enfold and run through the contemporary United Kingdom are reconfiguring, undergoing a period of rapid complex change.[2] The implications of these changes for the ways in which the ruling elite and the mass[3] of the population understand themselves are profound.[4] The political-cultural project[5] of Britain/Britishness is no longer tenable and will have to adjust to new circumstances, crucially, the intermingled dynamics of the internationalized global system,[6] the three major regions[7] and the unfolding project of the European Union. The elite can be expected to resist change, a defensive accommodation with no very clear destination-in-view. On the other hand, for the mass of the population, the 'ordinary English', the possibility emerges of an extensive progressive domestic reorganization.[8]

The demands currently placed upon the British elite are severe. It is difficult to see how they can evade the impact of these sweeping structural changes. The post-Second World War posture of the elite has been one of deference towards the USA and scepticism towards the European Union.[9] The consequent policy stance taken towards the development of the European Union has been one of 'dilute and delay'.[10] Yet it is difficult to see the EU, as the British elite might desire, modulating into a loose free-trade area subordinate to the USA. The German/French elites (and peoples) are wedded to the idea of the EU.[11] As ever, the British elite can play for time, however, they are unlikely to succeed for not only are the commitments of core mainland members well established, but the community itself is now institutionally elaborate, robust and deeply interlinked with the state-machineries of all the member countries.[12] In contrast the mass of the population might be expected to favour changes, recasting those received political cultural identities that have allocated them a firmly subaltern status. Yet the risks of imputing goals are obvious, so this analysis looks instead to the possibilities lodged in routine processes: 'thinking with England'.[13] It points to an anticipated process, that is, an unfolding, increasingly obvious and accelerating sequence of likely changes.[14] This anticipated process is grasped in terms of the metaphor of 'relocating England', the possible re-articulation of the political-cultural ideal of England/Englishness.

Identity and political-cultural identity

The disciplines of the social sciences offer a series of discussions of identity: in politics (engaged citizens); in economics (rational maximizers); in sociology (bearers of roles/status); and in cultural studies (inhabitants of sign systems).[15] In the classical European tradition[16] we find a concern to grasp the logics of the lives of people – the creative agents whose patterns of understanding shape their reactions to shifting structural circumstances – and a central role is made for ethnographic enquiry, detailing patterns of agent understanding.[17] An ethnographic-biographic approach[18] to identity enables the ways in which people read and react to enfolding structures to be unpacked schematically in terms of a trio of ideas: locale (the place where people live), networks (the ways in which people interact) and memory (the understandings which are sustained and recreated over time). These point to the ways in which (i) people inhabit particular places which are the sphere of routine activity and interaction and are richly suffused with meanings; which in turn are (ii) the base for dispersed networks of exchanges; all of which are (iii) brought together in the sphere of continually reworked[19] private/public memory. These strategies can be used at various levels – individual, group, organization, institution or states. The key is an appreciation of the fluid, relational, elaborate and social nature of identity.

The notion of political-cultural identity points to the way in which individuals and groups locate themselves with reference to the ordered political communities (polities) within which they take themselves to dwell; the way they handle questions of power and authority. The immediate locale will provide the person with a series of sources of practical knowledge: family, neighbourhood, organization, institution and media. Thereafter, agents form networks and lodge themselves in dispersed groups, and these groupings of persons order their relationships with other groups within the wider community. Finally, individuals and groups order their understandings of contingent, shifting patterns of power and authority in continually reworked memory. This can take a series of forms: the material of folk traditions, commonsense and local ideology (the material of little traditions); the histories of organizations and institutions; the ideas current in the public sphere and media; the official truths affirmed in the machineries of states; and ideas of nation[20] (the material of great traditions). In these various spheres we are looking at the production, dissemination and practical effect of sets of ideas about how the polity and its inhabitants are ordered and might develop (the range of possible/permissible lines of prospective action). Over time, there will be a series of elite political projects, each a way of reading and reacting to shifting circumstances, each shaping distinct forms of life and patterns of understanding.

The historical trajectory of the polities of the Isles

The Isles have been home to a series of polities, some located within the islands, others linked to the mainland.[21] It is a contingent sequence, marked by continuities, breaks and the production of novel forms-of-life and patterns of understanding. Any claim to an essential political continuity of the peoples of the Isles would be false. The available intellectual resources let us detail the contingent sequence of polities that have been present in the archipelago.[22] The history reveals a sequence of distinct polities which, in turn, provide intellectual, cultural and moral resources for contemporary reflection. Or, put another way, there have been several 'Englands' and several 'Britains' and these variously remembered historical episodes present us with a stock of resources – images of England and images of Britain – with which we can make preliminary sense of our present situation.

Norman Davies identifies three attempts at constructing England. The first is Offa (eighth century), then Alfred (tenth century) and both failed. The third, successful, attempt was the 1326–1603 exercise associated with the Tudors. Davies points to a series of factors: the separation of the polity from the various configurations of French feudal power (the Hundred Years War 1337–1453); the creation of 'middle English', that is, the language; the process of moving away from Roman Christendom (the Reformation); and the related strengthening of the absolutist Tudor state. The idea of England became the project, driven by the Tudor state, of the southern elite, and was popularized/ legitimized in a myth of England as an independent 'sovereign empire'[23] (later celebrated in the work of Shakespeare[24]). Davies speaks of 'the Englished Isles' (1326–1603). It is the crucial referent for an Anglo-centric history that obliterates the extensive interlinkages of the Isles and the mainland.[25] Thereafter, the Civil War of the 1640s saw an end to the absolute monarchy, power to the landowners and the rise of the towns.[26]

In the seventeenth and eighteenth centuries the English polity developed domestically as a progressive liberal commercial society,[27] a model for many subsequent enlightenment thinkers,[28] and internationally as extensive trade linkages were built up with mainland Europe and the colonies in North

America. Yet from the second half of the eighteenth century the ruling elite found themselves cut off from America (as a result of the American revolution in 1776) and mainland Europe (as a result of the French revolution in 1789). Linda Colley[29] argues that their response was to reinvent their polity through an aggressive outward expansion. An official nationalism was thus invented: Britain. It was an elite nationalism, not a popular one. A worldwide empire was accumulated,[30] and as imperial Britain was constructed the idea of England fell into the background.[31] The official nationalism was an elite construct: one might identify many ways in which top-down ideas and demands intersected with subaltern ideas and practices, but provisionally one might speak of an 'Anglo-British' identity.[32] It was the elite reworking of available ideas of England as a cultural residue lodged within the over-arching notion of Britain.[33] It was a confection shaped in part by the prejudices of an imported German royal family ('olde England' was emphatically *not* French[34]). It is arguably a key source of the present confusion of many in the majority population of the UK who are enjoined to run together Britain/England in a blurred mish-mash of imagery typically understood as 'heritage'.[35]

The nineteenth century saw the industrial revolution, free trade preached and the accumulation of a global colonial empire. The high tide of empire was attained in years before the Great War, 1914–18. Yet the war shattered the cultural confidence of European elites as the barbarism of industrialized warfare sat poorly with claims to the status of an advanced world civilization.[36] The Hohenzollern, Hapsburg and Romanov empires were destroyed and a problematical liberal state system established in Eastern and Central Europe. A series of failures followed: the economic depression, the failure of democracy and the rise of fascism.[37] The unhappy period culminated in the Second World War and any residual claims to civilized pre-eminence on the part of the Europeans were swept away. The period 1945–89/91 saw occupation, division and the construction of the apparatus of 'block-time'.[38] The British polity was absorbed into the American sphere. The business of recovery from war damage was begun, and a measure of success achieved.[39] But there was to be no recovery of the extensive pre-war imperial project. The political elite's residual aspirations, locating Britain at the centre of the intersecting spheres of the USA, Commonwealth and Europe, were fatally damaged by the 1956 Suez debacle. The British political elite accommodated to Europe,[40] disposed of the remnants of empire and reconciled themselves to US pre-eminence via the inter-related conceits/consolations of the special relationship, the role of number one ally in NATO and the over-arching idea of the English-speaking peoples. The British elite presented themselves as a bridge between America and Europe.[41] The strategy was essentially defensive, a minimalist response to changing circumstances; and from 1973 to 1989/91 it was sustainable. But the events of 1989/91 meant that the old familiar certainties were overthrown.

The project of the EU gained an unanticipated salience. In Britain, it became the central point around which politics revolved.[42] However, the period 1989/91–2003 was an interregnum. A series of interpretations were advanced – the end of history, the clash of civilizations and globalization – and all looked to fundamental continuity.[43] A wider crisis was precipitated by the 2003 Iraq war. It showed that a new post-cold war settlement/understanding was required. It also revealed a continuing British political elite concern for alliance with the USA.[44]

After 2003 – change, identity and action

The events of 1989/91–2003 swept away familiar patterns of understanding and the idea of Britain as one constituent part of a dispersed grouping of Western nation-states led by the benign hegemon of the USA is now untenable. This way of thinking about the circumstances of the polity does not grasp the dynamics of changing macro-structural contexts (where the demands of the internationalized global system, significant regionalization and the project of the European Union are crucial); it does not constitute an effective scheme of political-cultural identity (where the demands for democracy lodged within the modernist project now find wide expression within mainland Europe, the USA and, in rather different forms, in East Asia); and it does not offer any plausible routes to the future for either elite or mass (where a continuation of the status quo would leave the polity falling slowly behind the developmentally more advanced/progressive countries of Europe, America and East Asia). The structural pressures for change are strong. A process of Europeanization might be envisaged. It will be a contested process. The practical tasks are daunting; so too the related work of imagination. The elite may be expected to resist.[45] The masses have an opportunity. In any event new patterns of self-understanding are in prospect. It is in the context of new ideas of Europe that we might find new expressions of Englishness.

It is clear that formal and informal identity claims are lodged in developing time, and what one generation affirms as important, central and true, the next generation will rework. We inhabit tradition, but it is a continually reworked tradition. Ordinarily one might expect this reworking to be accomplished within the frame of an expectation of continuity, of things going on pretty much as they always have (even if the element of continuity involved a significant measure of myth), but in the present case the shift into Europe entails a disjunction, an acknowledgement of a new phase in the political-cultural development of the polity. In this context the contribution of the project of the European Union – contrived in response to complex patterns of global and regional structural change – is important; it offers a new political-cultural space and it is within this space that the idea of England might be reconsidered. One starting point is the available stocks of images; England can be characterized in a series of ways: (i) liberal, commercial and energetic;[46] (ii) rural, pacific and enduring;[47] and (iii) radical and forward looking.[48] More can be added: (iv) there is a multi-cultural England;[49] and (v) we have the Anglo-British orthodoxy, the core preoccupation of this text. These images provide resources. Any new idea of England/Englishness will not emerge quickly or straightforwardly; the process (practical/imaginative) will be complex, long drawn out and contested, hence the metaphor, relocating England. It unpacks in a number of ways: (i) relocation, as finding what was lost; (ii) relocation, as moving to a new place; and, more energetically, (iii) relocation, as reinventing and reconstructing within a new context.[50]

Notes

1 See S. Strange 1988, *States and Markets*, London, Pinter.

2 Here I have in mind in particular economic and political power structures: the rise of East Asia and the European Union as economic powers has been widely remarked upon, so too the shifting patterns of associated political power. I have benefited in particular from the work of Paul Hirst, Grahame Thompson, Richard Higgot, Robert Wade and Linda Weiss. These matters have entered the UK public sphere slowly, but as a rough time period, they became clear after 1989/91, and clearer after 2003. Over the longer term, significant dates include 1971 (Bretton Woods exchange rate system abandoned), 1978 (Deng Xiaoping inaugurated reforms in China) and 1985 (Plaza Accord revalued the yen, leading to rapid growth/bubble economy). I have pursued these matters in P. W. Preston 1997, *Political-Cultural Identity: Citizens and Nations in a Global Era*, London, Sage; P. W. Preston 1998, *Pacific Asia in the Global System*, Oxford, Blackwell.

3 D. Canadine 1998, *Class in Britain*, Harmondsworth, Penguin. Canadine argues that there are three political models through which the UK population has grasped the issue of class – hierarchical, triadic and dichotomous – and that sometimes one or other comes to the fore. At this time, however, I think the dichotomous model has a particular resonance, as the UK elite are clearly very uneasy about the prospect of the European Union. In this text I shall argue, as have others, that it represents a threat – that of 'democratization'.

4 A series of authors have addressed these matters – I have benefited in particular from the work of Linda Colley, Norman Davies, Tom Nairn, Patrick Wright, Richard Hoggart and Benedict Anderson and Ernest Gellner, all of whom are concerned, one way or another, with the ways in which elites and masses read their circumstances – that is, how they make sense of their (political) worlds.

5 Ideas of nation are not simple givens, they are constructs, contingent, in need of routine making and remaking – in this sense they are projects. Nations are also more than functional political apparatus – they offer packages of ideas and identities, hence the tag 'political-cultural project'. See Preston 1997.

6 P. Hirst and G. Thompson 1999 (2nd edn), *Globalization in Question*, Cambridge, Polity.

7 An interesting discussion of the EU was offered recently by B. Laffan, R. O'Donnell and M. Smith 2000, *Europe's Experimental Union: Rethinking Integration*, London, Routledge. They characterize the EU as a contingent accumulation of institutional structures, a 'deep regionalism'.

8 This book is concerned with the English who make up approximately 86 per cent of the population of the United Kingdom. It is concerned with how they understand themselves and with how the developing space of the European Union might impact upon those self-understandings. The book explores the possibilities for a re-articulation of the notion of Englishness, speculating about a 'European-Englishness'. The book is not concerned with Scotland, Wales or Northern Ireland. It is not about devolution. It is not a history of the United Kingdom. It is a book about changing political-cultural identities and projects. As the project of 'Britain' has passed its sell-by date, the book considers the future of the English; others will have to concern themselves with the remaining 14 per cent – the Scots, the Welsh, and the people who live in Northern Ireland.

9 British elites have looked to the model of the European Free Trade Association (EFTA), a loose free-trade area. There is confusion about the UK's 'place in the world', for example with Suez in 1956, then EFTA in 1959, and Macmillan's application to join the EEC in 1961. See D. Urwin 1997, *A Political History of Western Europe Since 1945*, London, Longman, pp. 107–11, 166–73.

10 P. W. Preston 1994, *Europe, Democracy and the Dissolution of Britain*, Aldershot, Dartmouth.

11 In order to keep the detail manageable, in speaking of the European Union the text will focus on a trio of countries: France, Germany and the UK. The shifting attitudes of these and other (and newer) elites/masses have been (and are) extensively canvassed. On this see M. Marcussen *et al.* 2001, 'Constructing Europe? The Evolution of Nation-State Identities' in T. Christiansen *et al.* (eds), *The Social Construction of Europe*, London, Sage. Marcussen argues that the political elites of France and Germany embraced the EU project: France after a series domestic political crises centred on its image/place in the world; Germany as a reaction to the catastrophe of fascism/war. In contrast the UK elite never embraced the project, as it implies great domestic reforms at which the elite baulked.

12 Laffan *et al.* 2000.

13 R. Colls 2002, *Identities of England*, Oxford University Press, chapter 22.

14 C. W. Mills 1970, *The Sociological Imagination*, Harmondsworth, Penguin. Mills remarks that it is the mix of public/private that sparks the scholarly imagination. Thus the confidence is social scientific, that is, it flows from an interpretive-critical analysis of extant patterns of structural change as they enfold the UK polity, but the interest is also personal, that is, the critical aspect identifies a route towards a goal taken as desirable, i.e. the democratization of the UK polity.

15 All pursued at greater length in Preston 1997.

16 The text is located within what I think of as the classical European tradition of social theorizing. This is a quite particular view of the nature of social theorizing.

In *formal terms* the text is interpretive-critical. The business of social theorizing is understood to be generic to humankind, the business of making sense of the social world. It is diverse in its forms. It finds expression in great traditions, little traditions and the routines of everyday life. One formal strategy of making sense is available in the work of the social sciences. The social sciences themselves are diverse – there are identifiable national traditions, disciplinary frameworks and strategies for putting the resources of the social sciences to work in practical analysis. Any exercise in social scientific work can be characterized as a 'mode of social theoretic engagement' – a drawing down on available intellectual stocks in order to make arguments designed to inform the practice of particular audiences. The arguments made will entail a spread of commitments: epistemic, ontological, methodological, ethical and practical (arguments entail action). The criteria of success (what is a better/worse argument) will be found in judgements of the coherence, plausibility and utility of the arguments presented (as the exchange of context, theorist/theorizing and audience unfolds the upshot of a successful episode will be 'better ways of understanding' (for audience and theorists)). The classical European tradition of social theorizing is concerned, *more substantively*, with the emancipatory elucidation of the dynamics of complex change in the ongoing shift to the modern world. The tradition is concerned with analysing structural change and agent response, including, quite centrally, the activities of political elites as they fashion, legitimate and deploy political-cultural projects designed to plot routes to the future for the populations they control/lead. Amongst these concerns, one area of enquiry looking at patterns of understanding focuses on the political sociology of the response of agents to complex change – the ways in which groups read, react and respond to the demands placed upon them by rapid complex changes in their social environments. All this is furyther pursued in P. W. Preston 1994, *Discourses of Development: State, Market and Polity in the Analysis of Complex Change*, Aldershot, Avebury.

17 A. Giddens 1976, *New Rules of Sociological Method*, London, Hutchinson; A. Giddens 1984, *The Constitution of Society*, Cambridge, Polity.

18 Preston 1997.

19 Memory is active: see Raphael Samuel 1994, *Theatres of Memory Vol. 1: Past and Present in Contemporary Culture*, London, Verso.

20 In this text read as constructed (see E. Gellner 1983, *Nations and Nationalism*, Oxford, Blackwell), as a top-down official national identity (see B. Anderson 1983, *Imagined Communities*, London, Verso), and as one more layer of political-cultural identity, not, that is, an evident master status.

21 N. Davies 2000, *The Isles: A History*, London, Papermac.

22 A position I take from Davies, Colley and Nairn. An alternate line stresses deep continuity – see R. Scruton 2001, *England: An Elegy*, London, Pimlico; J. C. D. Clark 2003, *Our Shadowed Present: Modernism, Postmodernism and History*, London, Atlantic Books.

23 Davies 2000, p. 388.

24 Davies 2000, p. 426.

25 This is a complex debate pursued by historians: see Davies 2000; also E. Jones 2003, *The English Nation: The Great Myth*, Stroud, Sutton Publishing. In brief, the Tudor period is read by later historians as the key to an exceptionalist history which separates the polities of the Isles from those of the mainland; it is an ideological reading in other words.

26 Again, this is a matter of debate amongst historians – I owe this view to Christopher Hill. Davies (2000) rejects it in favour of the inter-relations of the three nations of the Isles plus religion. Tom Nairn (1988, *The Enchanted Glass*, London, Hutchinson) points out that the new polity looked back to the Renaissance city-states and the Dutch Staadholderate for models, not forward to the modern world of the enlightenment.

27 Roy Porter's point. See R. Porter 2000, *Enlightenment: Britain and the Creation of the Modern World*, London, Allen Lane.

28 See for example I. Buruma 1999, *Voltaire's Coconuts, or Anglomania in Europe*, London, Weidenfeld.

29 L. Colley 1992, *Britons: Forging the Nation 1707–1837*, Yale University Press.

30 In the periphery – north America – they spoke of themselves as Britons. See C. Bridge and K. Fedorowich (eds) 2003, *The British World: Diaspora, Culture and Identity*, London, Frank Cass.

31 R. Weight 2002, *Patriots: National Identity in Britain 1940–2000*, London, Pan. In the Introduction Weight argues that after the Act of Union the English were rather unenthusiastic about the idea of Britain, as their identity as English was subsumed within a new official idea of British. K. Kumar (2003, *The Making of English National Identity*, Cambridge University Press) makes a similar point. Interestingly, Colls (2002) argues that the identity – English – was played down by an elite nervous of an English nationalism (pp. 43, 51). In any event, Englishness flowed out of the routine experience of the English. Weight uses the term 'Anglo-British' which seems to gesture to what was put in its place. Interestingly, Scruton (2001) remarks that England was 'home', *as seen from overseas* by those serving in the empire (pp. 2–3).

32 Weight 2002.

33 The idea of 'Englishness' comes later (first used in 1805), and reflects the influences of German romanticism. The term grasps the 'essence of nationality'. P. Langford 2000, *Englishness Identified: Manners and Character 1650–1850*, Oxford University Press, pp. 1–2.

34 The non-French nature of standard Anglo-British Englishness is noted by Norman Davies (2000). The nineteenth century invention of 'olde England' is discussed by Langford 2000, chapter 1.

35 P. Wright 1985, *On Living in an Old Country*, London, Verso.

36 Eric Hobsbawm refers to 1917–89 as the short twentieth century, yet the history of the UK and Europe in this period falls more clearly into two inter-related phases. It is not the rise and fall of the European state-communist experiment that is crucial, rather the long drawn out European civil war of 1914–45 with its subsequent period of division and occupation (1945–2003). E. Hobsbawm 1994, *The Age of Extremes: The Short Twentieth Century 1914–1991*, London, Michael Joseph.

37 Mark Mazower 1998, *Dark Continent: Europe's Twentieth Century*, New York, Alfred Knopf.
38 E. P. Thompson's reference to the institutions, ideas and popular politics of the cold war.
39 Thus the whole welfare state. However, much debate begins at this point as a 'declinist' literature develops.
40 Abandoning EFTA and joining the EEC – at the third attempt – in 1973.
41 A contemporary discussion is offered by A. Gamble 2003, *Between Europe and America: The Future of British Politics*, London, Palgrave.
42 Prime Minister Margaret Thatcher was sacked, her successor John Major signed the Maastricht Treaty and later Tony Blair offered further positive mood music about the UK and the EU.
43 On this, broadly, see Z. Laidi (1998, *A World Without Meaning: The Crisis of Meaning in International Politics*, London, Routledge) for a diagnosis offered from 'inside the interregnum'.
44 The central theme of Gamble 2003.
45 H. Young 1999, *This Blessed Plot: Britain and Europe from Churchill to Blair*, London, Papermac.
46 See Porter (2000) on what he calls anachronistically the 'British enlightenment', the key to which was energetic opportunistic commerce, a vulgar, vital social style.
47 Evidenced in novels of Thomas Hardy, celebrated by George Orwell, and recently celebrated by Scruton 2001.
48 The Levellers, Tom Paine and the Chartists; see E. P. Thompson *et al.* in H. Kaye 1984, *The British Marxist Historians*, Cambridge, Polity.
49 Y. Alibai-Brown 2001, *Who Do We Think We Are: Imagining the New Britain*, Harmondsworth, Penguin.
50 'England/Englishness' is usually thought of as the province of the political right/conservatives – thus 'little Englanders' or 'English football hooligans' – but one can point to various strands – rural, urban, radical and multi-cultural. Any 'relocation' will be contested (not (just) nostalgic). Any new notion of 'England' would emerge from the process of relocation; there is no ready made or essential model. One might also argue that if the idea/cultural resource of 'Englishness' is neglected, it will be the right/conservatives who will use it. See Gunter Grass (2002, *Crabwalk*, London, Faber and Faber), who argues that the German political left neglected the suffering of the German people at the end of the Second World War and thus left the territory to be occupied by the political right.

2 Political-cultural identity[1]

The classical European tradition of social theorizing is centrally concerned with elucidating the dynamics of complex change in the ongoing process of the shift to the modern world.[2] It is an intellectually diverse tradition that finds practical expression in various discrete modes of engagement. The tradition has a characteristic set of preoccupations and generates a rich vocabulary. It begins with a concern for the detail of the social processes, the heartland of social science. The wider historical-structural dynamics are grasped in terms of the ideas of conjunctures, disjunctures, phases and trajectories. The tradition has developed over time in line with internal debates and through a wide spread of engagements with the wider social world.

In substantive terms, enquiry revolves around the view that it is within changing structural circumstances that contingent patterns of social life develop. First, the population of any territory (or country) will generate over time a distinctive historical trajectory through the routines of ordinary practical activity and institution building. It is a contingent accumulative process. The accumulated cultural capital underpins the present community, itself in the process of dissolving into the future.[3] Second, the historical trajectory will not be smooth. The trajectory we point to is the present record of a contingent process, the stringing together of a series of unfolding events. There are recurrent concerns: thus peoples' concerns with their families, livelihoods and communities. Patterns can be identified retrospectively, but there are no unfolding processes, no teleology, nor deep mechanisms. The trajecto ries will be discontinuous. A concern for macro historical-structural analysis implies a concern for conjunctures, the ways in which a given territory (country) is located within wider patterns of power. Third, the historical trajectory will evidence phases; that is periods of stable economic, social and cultural forms-of-life. Of course there is a related concern for disjunctures (the episodes of rapid complex change which re-direct development trajectories) –

these can be generated either internally (economic or social breakdowns or novel patterns, finding expression in new political projects), or externally, as changing structural circumstances demand creative agent responses. And finally, the patterns of life of the people of a territory, or country, will be ordered in systems of institutions, routinized patterns of life and formal organizations. It will be a contingent pattern. The political elites are embedded within these enfolding structural patterns and processes. It is from within the context of definite circumstances that they will endeavour to read and react to change, formulating political-cultural projects oriented to the future and seeking to mobilize their populations accordingly.

The notion of identity lets us examine these matters further. Identity expresses the response of the agent to the demands of the enfolding structures, a balance between personal subjective experience and the demands carried in wider social processes and structures. Identity is not fixed, it is dynamic, an exchange between person and social processes. The notion of political-cultural identity points to the way in which persons lodge themselves within enfolding social processes understood as ordered communities. The dynamic is the same: the exchange between person and wider, more powerful social processes. The business of political-cultural identity can be approached in terms of sets of ideas, from the way in which we think of ourselves, to the sets of ideas carried in organizations, institutions, wider communities and general 'great' traditions. Once again, any personal identity is dynamic, a balance between alternate ideas, a way of reading enfolding structural circumstances. If the structural circumstances enfolding the life of a community change, then so too – necessarily[4] – will the identities of that community. As the circumstances that sustained the political-cultural project of 'Britain' change, so too will the project. In these changes we find the possibility of 'relocating England'.

An ethnographic/biographical approach to identity

The materials of the social sciences find practical expression in a variety of discrete modes of social theoretic engagement, ways of making arguments on behalf of particular audiences. Any exercise in social theorizing will involve commitments about intellectual procedures and the goals to which enquiry is oriented.[5] The theorists of the nineteenth century variously affirmed ideals of rationally secured social progress. Bauman[6] distinguishes 'interpreters' and 'legislators', with the former oriented to discourse within the public sphere, and the later concerned with the provision of technical knowledge within an increasingly rationalized social world. In the late nineteenth century, for a variety of reasons,[7] a spread of restricted technical, professional, discipline-bound knowledges emerged. The process of disciplinization is an elaborate continuing social process where each discipline lays claim to a particular object sphere, a relevant method, a body of accumulated wisdom and the status of a profession within the knowledge marketplace. A fragmented intellectual pattern results. There are, consequently, a multiplicity of ways of grasping the ideas of identity and political-cultural identity. The ways in which orthodox economists, political theorists, sociologists and cultural analysts approach the notion of identity can be grasped in outline in terms of their fundamental logics of enquiry. In general terms each of these four approaches offers a particular route into the issue of identity and thereafter the matter of political-cultural identity: (i) market and consumption (looking at choice, competition and life-style); (ii) states and ideologies (looking at power, citizenship and legitimation); (iii) society and socialization (looking at relationships, structures and learning); and (iv) culture and tradition (looking at practices, ideas and involvement).[8] In this text an alternative strategy informed by the classical European tradition of social theorizing is used.

The central concern is with the richness, specificity and changing nature of identity. The contingency, immediacy and fluidity of identity can be grasped, whilst at the same time granting its routinely un-problematical nature; thus, most people, most of the time, find 'identity' wholly un-problematical. A start-point for enquiry can be found in a simple strategy of formal reflection. If we begin with an idea of identity as the way in which people,[9] more or less self-consciously, locate themselves in their social worlds we can review the various aspects of identity in the fashion of a substantive 'ethnographic/ biographic' report.[10] Identity can be unpacked in terms of the ideas of locale, network and memory. This trio points to: (i) the ways in which we inhabit a particular place which is the sphere of routine activity and interaction, and is richly suffused with meanings; (ii) which, in turn, is the base for a dispersed spread of networks of exchange with others, each centred on particular interests/concerns; (iii) all of which are brought together in the active processes of continually reworked memory. The approach generates a series of aspects of identity, which can be taken to be produced by an ideal-typical agent in response to a naïve question – who are you? The answers given in this simple strategy generate a rich agenda.

A sense of locale

The notion of locale points to the immediate sphere of practical activity within which we move. It will involve a specific group of people. It will be ordered around a set of routine practices. It will involve a series of familiar and regularly used locations. It will involve a taken-for-granted yet richly known background, a place. All this we can summarize as locale. The locale is the immediate source of personal identity. It is the sphere of routine activity and interaction and is richly suffused with meanings.

The sphere of family

The family is central to the construction and nature of identity. In brief, if we are asked our identity ('who are you?'), the first reaction is that we give our family name. '*My name is X*.' As the first element of identity the family is the location of those routine social practices that shape our fundamental selves: physical self, voice and the social habits of the domestic sphere. It is the location of our memories of childhood, of parents, siblings and relatives. Hilary Mantel grasps the depth of its reach:[11]

> For a long time I felt as if someone else were writing my life ... About the time I reached mid-life, I began to understand why this was. The book of me was indeed being written by other people: by my parents, by the child I once was ... [And] I began this writing in an attempt to seize the copyright in myself.

The family is not merely our social location, family locates us – fixes us in place in the community. Richard Hoggart, writing about his own childhood in 1930s Leeds, grasps this beautifully:[12]

> That indicates yet again the main pattern of our lives ... We had virtually no lines out to lives, interests, concerns, beyond ourselves. This was not innate selfishness or self-absorption; these were the terms ... forced on us by the stringency with which our mother had to operate. School friendships stopped at the playground gates, since there could be no exchange of visits; we were the only kids in our yard so no one else played there. Our only visits were to our paternal grandmother in Hunslet. We belonged to no sporting or recreational or community or school groups; we were wholly outsiders because we had to be so much insiders and, since we know no other way, we did not seek to belong.

In a more lyrical vein, from English literature[13] directly, showing family and place linked in memory, Laurie Lee recalls his Cotswolds childhood in the last years of the Great War:[14]

> I was set down from the carrier's cart at the age of three; and there with a sense of bewilderment and terror my life in the village began. The June grass, amongst which I stood, was taller than I was, and I wept ... For the first time in my life I was out of sight of humans ... From this daytime nightmare I was awakened, as from many another, by the appearance of my sisters ... Marjorie, the eldest, lifted me into her long brown hair, and ran me jogging down the path and through the

> steep rose-filled garden, and set me down on the cottage doorstep, which was our home, though I couldn't believe it. That was the day we came to the village, in the summer of the last year of the First World War. To a cottage that stood in a half-acre of garden on a steep bank above a lake; a cottage with three floors and a cellar and a treasure in the walls, with a pump and apple trees, syringia and strawberries, rooks in the chimneys, frogs in the cellar, mushrooms on the ceiling, and all for three and sixpence a week.

In the more austere language of the social sciences, the family has been a routine topic for political, policy and intellectual debate about the nature and direction of society. The analysis of the family has been widely pursed: the economic roles of the family are numerous; the family in anthropology/sociology shows various forms; the family figures in psychology; and the family appears in various social and moral guises (with questions as to its centrality or the necessity of its radical reconstruction). The family has figured in ideological dispute about the nature and direction of society since the early nineteenth century. Friedrich Engels' work on the condition of the English working classes was the first in a line of social scientific analyses of the negative aspects of rapid unplanned industrialization, ameliorist work which runs down to the present day. And as the family locates us within society, it has been routinely discussed in terms of 'property'. The familiar rhetorical linkages of 'family' and 'property' offer a very particular idea of family as an historically specific linkage of human and social reproduction is generalized. The model extends into justificatory ideologies of community, continuity and tradition. Evelyn Waugh offers an elegiac celebration of class privilege, property and loss in *Brideshead Revisited.* These themes find expression in party politics – the rhetorics of 'tradition' and 'continuity', in turn condemned in terms of the demands of 'social equity' and 'progress'. Yet these familiar rhetorical confections act to block the fuller appreciation of the richness, diversities and commonalties of the experience of family. When we reflect upon our selves, the stock of memories takes us to family-in-locale.

The neighbourhood, or local area

The notion of locale encompasses routine activity and understanding. It encompasses the immediate sphere within which the family is located. The 'local area' comprises the street (with the neighbours), the local shop, the school and maybe the pub. It might be urban, or sub-urban or the scattered territory of rural areas. In brief, if we are asked our identity one reaction is to tell the questioner where we live. '*I live in Y.*' The local area is intimately present in our identity. Again, it is present in voice, or physical manner, or in dress. The local area frames our memories of childhood, of parents, siblings, relatives, neighbours and the general spread of figures within the neighbourhood. Mantel recalls her school:[15]

> My last years at primary school had been conducted under the eye of Mother Malachy, which had rolled at me around the curve of her starched headdress … They were long, those days … [and the] lights burned all day in Hadfield winters, the great radiators puffed and fumed and stank, the odour of Wellington boots and nit lotion and nun became so thick you felt you could graze it with your knuckles; you were, very often, spoiling for a fight.

In a similar vein, so does Lorna Sage:[16]

> Back there and then in our childhoods … in the late Forties, Mr Palmer seemed omniscient. He ruled over a little world where conformity, bafflement, fear and furtive defiance were the orders of the day. Every child's ambition at Hanmer school was to avoid attracting his attention … We all played dumb, the one lesson everyone learned. We'd have seemed a lumpen lot: sullen, unresponsive, cowed, shy or giggly in the presence of grown-ups. A bunch of nose-pickers and nail-biters, with scabbed knees, warts, chapped skin and unbrushed teeth. We shared a certain family resemblance, in other words.

Richard Hoggart captures the broader social map of the local area of his childhood days:[17]

> Back home, in the streets again, the big physical markers … were the pubs, the chapels and the public library. .. The life of those streets, was, then, a matter of multiple fine-gradings … This cluster of streets belonged to the Irish, this thought itself slightly better than the others around it, and this was thought to be slightly less respectable. Or the divisions came at the main road or on each side of a medium-sized road which took off at right-angles to the main road; or the walls of a small factory marked the boundaries. Superimposed on those were the divisions indicated by the regular worshipers and customers at the chapels or pubs scattered throughout, to some extent cross-cutting other divisions. That was 'our' chapel or 'our' pub, those others 'theirs'. That fine shading by possessive identification extended to all the local shops: he was 'our' butcher, grocer, confectioner; and we were quite sure our shops were better than 'theirs'.

All these experiences serve to identify our background, and in particular, class. These matters have concerned social scientists and writers. In the former case: (i) the economic roles pursued in the local area will tend fall within 'the informal sector', for example, as we perform small services, or help our neighbours; (ii) in anthropology/sociology the local area points to the sets of various and detailed local rules which run through people's lives, the ethical resources of 'little traditions', the moral codes embedded in routine practice – thus we exchange favours, or in the pub, affirming an ethic of reciprocity, we 'buy our round'; and (iii) a local sphere of judgement can be identified – folk knowledges and common sense – for example the way in which a local area will be characterized, variously, as a good area, or as a rough one. In the latter case there are many examples, thus D. H. Lawrence's Nottinghamshire in *Sons and Lovers* (1913), Kingsley Amis's South Wales in *The Old Devils* (1987), John Fowles' Dorset in *Daniel Martin* (1977), Hanif Kureishi's London in *The Buddha of Suburbia* (1990), or Monica Ali's Tower Hamlets in *Brick Lane* (2003).

Substantively, it could be argued that the local area in the UK has been set to one side in the shift to the modern world and its subsequent development. The shift to the modern world entailed extensive economic, social and cultural changes – all of which impact upon local areas. A familiar characterization of the local area would make it passive, without power, subject to the impact of macro structural changes and little more than an adjunct to the inevitable activity of the domestic sphere. However, the local area is home to many social activities and networks. When we reflect upon our selves the stock of memories relates back to family-in-locale where the most immediate sphere of that locale was the local neighbourhood.

The wider local area, or community

As we move outwards from the family we shift to the local neighbourhood and thence to a slightly wider grouping where we operate as members of the wider local area or community. Hoggart, reflecting on the depredations of the politics of the 1980s, remarks:[18]

> Neighbourliness survives amongst middle-class as much as in working-class people. It is narrow in the sense of being confined to a restricted neighbourhood. Within that area it is all-pervasive ... It is also the single most sustaining communal practice in English society ...

A community will have symbolic locations such as the church, the police station, the factory and the town hall. In brief, if we are asked our identity, one reaction is to tell the questioner where we are from. '*I am from Z.*'

As an element of identity, the local area, the community, is a further location for our memories of childhood: parents, siblings, relatives, neighbours, the important figures within the neighbourhood, and more broadly the features of the wider town or city.[19] The analysis of the community has been widely pursed in social science: the economic roles pursued in the community will be many, formal and informal; the community has been a key issue in anthropology/sociology and a focus for discussions of belongingness and the problems of rapid change; and in the area of ideology the community has long been contested territory, often cast in terms of rural versus urban. In the post-war years, Wilmott and Young offered a classic celebration of the ideal of neighbourhood and community in *Family and Kinship in East London* (1957). Numerous other studies have been undertaken with, more recently, Patrick Wright's *Journey through the Ruins* (1992) doing something similar. Wright captures the essence of the corner of East London in which he lives:[20]

> Among the under-estimated attractions of Dalston Junction is a street corner full of forgotten municipal services. The public lavatories are of the attended Victorian variety with wrought-iron railings, descending steps and lunettes in the pavement; and, on good days at least, they still function as originally intended - unlike many of their equivalents in more right-thinking and fortunately placed London boroughs that have been sold into private use as wine bars, pool halls, and design consultancies. There's a distinctly village-like Public Notice Board provided by the council but now only used by the fly-posting militants of the Revolutionary Communist Party and the exiled Turkish Communist Party. Proudest of all, however, is the 'Hackney Town Guide', which offers to orientate the enquirer with an apparently unambiguous message: 'You are here'.

In the sphere of arts and letters we can find diverse celebrations. The official ideology finds one expression in the ethical/aesthetic territory of the long-running radio series, *The Archers*. In this sphere the social world is subjected to what might be called 'BBC classic serialization' where social life within a community is assimilated to an hierarchical, harmonious and civilized unity in quiet decency. We find the official ideology expressed in the self-regarding images of the social and moral community of the elite – evidenced in the work of Iris Murdoch, dealing with the moral dilemmas of the fastidious English bourgeoisie. There are also counter-statements, the community of the oppositional/subaltern masses such as John Braine's *Room at the Top* (1959), Stan Barstow's *A Kind of Loving* (1960), or, more recently, in the work of those with migrant family backgrounds, thus Zadie Smith's *White Teeth* (2001). Also, moving in its own territory, is the sentimental community of the ordinary people captured in popular music – The Beatles' *Penny Lane*, or The Kinks' *Waterloo Sunset*.

Substantively we can say that the community in the UK has been read in quite different ways. On the one hand, it has been taken socially and ethically to be under pressure in the shift to the modern world, where the shift is a matter of cultural, political and economic changes that impact upon the community which thereafter has to resist in order to maintain its original and intrinsic vigour (a conservative story). On the other hand, in terms of modern politics the community of a polity (a rational republic) and the imagined community of nation have both been strongly advocated (a progressive story). Thus, when we reflect upon our selves the stock of memories relates back to family, neighbourhood and local area, or community.

The institutional aspects of locale

The formal organizations within which people routinely operate offer a further aspect of identity. These organizations are removed from family and immediate locality. They may be voluntary (church, community group, sports club and so on), or commercial (the Automobile Association, or airline frequent flyer schemes where one becomes a 'member'), or professional (with associations or unions). The key organization, for most people, will be the place of work. The analysis of work has been widely pursed in social science: the economic roles pursued in work will be many; the nature of work has been a key issue in anthropology/sociology and the locus for discussions of alienation/anomie; and in the area of ideology the nature of work has been contested territory, often cast in terms of exploitation versus necessity. The shift to the modern world was a matter of cultural, political and economic changes which included centrally the sphere of work: the shift from traditional craftwork to industrial work.

In brief, if we are asked our identity one reaction is to tell the questioner what we do for a living. '*I am an X by trade.*' There is a wealth of material here. Scholarly work has been shaped by reformist political agendas, from Friedrich Engels to Beatrice Cambell retracing George Orwell's steps in *Wigan Pier Revisited* (1984). There are also many memoirs. Keith Waterhouse writes of his first job:[21]

> Consulting the list of available vacancies at the College of Commerce, I found what promised to be the ideal situation ... the firm of J. T. Buckton & Sons, the Leeds Funeral Furnishers ... The small office was in a converted shop ... The office was divided, as required, into cubicles and cubby-holes ... my duties ... proved remarkably light and extremely congenial. My first task of the day was to take down the heavy window shutters ... Then I had to go across the street to the Craven Dairies cake shop and cafe with a filing tray of mugs for our early morning coffee. There were eight of these, the office being absurdly overstaffed ... I was put in charge of the postage book. This suited me perfectly, for much of Buckton's correspondence was with other firms in Town, and it was the partners' policy that any letters within the Leeds 1 postal district had to be delivered by hand ... From the start, then, I enjoyed a carefree forty-five minutes or so each morning ... There were other regular errands from which there seemed to be no pressing need to hurry back: collecting inscribed coffin plates from the engraving firm ... or placing death notices and property column small adds with the Yorkshire Evening Post ... Back at the office, life went on at a leisurely pace. A great deal of Craven Dairies coffee was drunk and such work as there was to do was performed against a background of chit-chat and badinage.

In UK common culture work figures in a series of ambiguous ways: the elite model is of the non-work of the supervision of landed estates, or the putatively semi-work of professions. The realm of modern industry generally has low status. In this sphere people wear 'blue collars' and 'get their hands dirty', and are of little account. The familiar spread of images of work – and thus the available clue to its reading by those who undertake it – do change over time as old occupations disappear and new ones emerge and are taken into the schedule of good/bad jobs. As regards the detail of these experiences, we can find many examples in English literature. There is a mythology of working-class work, such as Robert Tressel's *The Ragged Trousered Philanthropist* (1955), D. H. Lawrence's miners, or Alan Sillitoe's *Saturday Night and Sunday Morning* (1958). There is a bigger mythology of ruling-class non-work as found in Evelyn Waugh's novels. And there is John LeCarre's world of spies, which deals with the codes of honour and routines of self-deceit amongst the respectable professional servant classes (*Tinker, Tailor, Soldier, Spy*, 1974).

Locale and little tradition

The local sphere generates its own patterns of understanding – folk knowledges, local ethics and so on – the realm of little tradition, the ideas/ethics of subaltern groups. It is a matter of lives pursued within restricted spheres. In brief, if we are asked our identity one reaction is to tell the questioner how people like us think and act. One can say '*I am a Y by background*'. These distinctions are marked easily, as the judgements are built into our routines. Lorna Sage recalls the reaction to an unplanned pregnancy:[22]

> My mother came upstairs and opened the door, her face red and puffed up with outrage, her eyes blazing with tears … For a minute she says nothing and then it comes out in a wail, *What have you done to me?* Over and over again. I've spoiled everything, now this house will be a shameful place … I've soiled and insulted her with my promiscuity, my sly, grubby lusts … I've done it now, I've made my mother pregnant.

A little later, married, she looks for contraception:[23]

> Next I went to ask Dr Clayton about contraception … all he said was, 'Now that you're married your husband will take care of that'. What he was saying … was that he wouldn't aid and abet me in acquiring any control over my own fertility. In any case, he must have thought, I was now in all probability going to revert to white-trash type and have more babies, and in a way decorum demanded that I should; I was some sort of nymphomaniac, and mustn't be allowed to have my cake and eat it.

These social rules are deeply inscribed in routine practice. The distinctions pervade social life. They are suffused with judgements of class. Richard Hoggart, recalling the fine social gradations of his Leeds childhood, offers a contemporary example:[24]

> Years later, in the days when university external examiners were given first-class rail fares, I travelled from London to Brighton on the Pullman … After only a few minutes the chocolate-and-fawn, tight-uniformed and tight-arsed attendant assigned to that car came to me and said: 'Excuse me sir, but you do realise this is a first-class carriage?' I resisted the temptation to do a Lucky Jim, and said: 'Yes, and I've got a first-class ticket'.

In English literature, there are many examples of regional novels: those of Thomas Hardy, D. H. Lawrence, Raymond Williams, or more recently Melvyn Bragg. In all these cases the author locates the work in a clearly specified place, a distinctive region. It might be argued that such work is increasingly retrospective. These regional particularisms have been submerged within an Anglo-British common culture dominated by the southeast. The popular residues are submerged within a pervasive acquiescent little tradition – deference, consumption and individualism.[25] In the social sciences this sphere belongs to cultural studies, local history and the anthropology of the contemporary UK.[26]

Institutional truths

The extensive sphere of state institutions impacts upon identity. As persons move out of the informally structured sphere of family, locality and community they encounter formal organizations, as noted, and thereafter the immediate presences in their locale of the formal institutions of the state: the tax man, social security officials, health administrators, police, the officials of the local authority, and so on. All these formal institutions affirm institutional truths to which persons in exchange with them have to submit one way or another. In brief, if we are asked our identity one reaction is to tell the questioner our ID number. '*I am Z in your files.*'

The classical tradition of European social theory has made the ordered nature of modernity a key concern. The ideal of progress has been intimately linked with the achievement of rational order. However, there is a critical line, for example in Max Weber, an early analyst of the ambiguous model of 'rational bureaucracy'. The central claim of such theorists is that sets of values/identities are inscribed in the routines of bureaucratically ordered social spheres – in brief, the claims to neutral technical expertise are fraudulent.[27] None the less the demands of the bureaucrats bear down upon the population. In English literature there are available examples, as with John LeCarre's discussion of the life of an embassy in *A Small Town in Germany* (1969). In sum, when we reflect upon our selves the stock of memories will automatically vehicle the resources of those institutional locations within which we live and work.

Great tradition and official ideology

The 'great tradition' and 'official ideology' are spheres carrying cultural models of identity.[28] The formal institutions of the state affirm institutional truths to which persons are required to submit. They also embody broader sets of ideas, the official truths that the state affirms. In brief, if we are asked our identity one reaction is to tell the questioner our socio-political status. '*I am a member of an X society.*'

In sum, if we are asked who we are, then one obvious line of reply will draw on available delimited-formal ideological constructions, in this case ideas which designate the political group to which we take ourselves to belong, class or ethnic group or nation.

The media

Resonating with all these levels we have the realm of media-carried identity, centring on newspapers, magazines, advertising, radio and television. The exchange of the media with the social world upon which it comments and whose comments impact upon that social world is diverse: (i) the routine social, as in the output of light entertainment, which ties identities back into patterns of common-sense reflection upon the social world (think of a soap opera); (ii) the routine economic, in the form of consumer-oriented material, which ties identities back into given political-economic structures (think of advertisements or programmes devoted to consumer activities – food, cars, holidays, etc.); and (iii) the routine political, in the form of news and current affairs, which ties identity back into given political institutional structures and understandings (think of news programmes with their standard agendas). In brief, if we are asked our identity one reaction is to tell the questioner what we read or watch. '*I am a Z reader/viewer.*'

Knowledge of networks

The notion of network points to the wider spread of contacts that people will use. The idea of networks adds a spatial element to the business of identity. There will be a spread of practical activities, and these contacts could be professional, business, pleasure, or maybe family (as with emigration). The network is an element of our routine lives – in this sense it is present in our locale, it is not removed or remote, it is not 'elsewhere', it runs through our ordinary life. Thus the network carries further aspects of a person's identity as we invest our attention in the domestic sphere, thereafter the world of work, then the formal institutional sphere, and as a loose spread of ideas throughout the whole ensemble the world of the media, simultaneously present in the detail of life yet superficially distant.

The family sphere is the most immediate location for those sets of relationships that constitute identity. An elaborate network is build around it. The family may be geographically close or more dispersed. A young nuclear family will be concentrated but an older family with children away from home will be dispersed. The dispersal may be relatively restricted (in the same town) or quite distant (another town, or another country). The extended family of aunts and uncles may be widely distributed.

The world of formal organizations provides further dense networks. Here we encounter social practices that generate further aspects of identity – suppliers, buyers, customers, colleagues, officials and so on. It is a dense and extensive network, that of people routinely involved in the work of shop, factory or office – not physically present, but an indispensable part of the activity nonetheless.

The spatial nature of the networks that run through ordinary routines gives rise to the experience of boundaries, the points at which locales/networks stop. These boundaries can be crossed. A key step in altering the configuration of the basic locale, a way of stepping outside and joining in the wider world, thus making networks, is the physical and class move of entry to school. Hoggart grasps this clearly:[29]

> As you pushed off, out of the gravitational pull of your own small ring of streets through others you knew less and less well as the rings widened outwards, the associations, the known places, the accretions of memory all thinned out: you were enclosed within yourself, looking nervously but doggedly ahead, one person going through what would have seemed to outsiders like featureless, dull streets to a new order of life, new words, new ways of behaving, new kinds of excitement, a novel sense of possible openings-out, 'prospects'. The first day at university, the first walk into a barrack-room where you will sleep as a recruit, are similar nervy beginnings if that is your disposition; neither has as much force as that first walk to the grammar school and entry to your first classroom and all the new faces, at eleven.

The little tradition that will encompass these activities and understandings will include a reference to the territory, routes and boundaries that make up the particularity of the place. In answer to the question 'who are you?' we can reply

'I am from the north', or the south-west, or wherever. Such an answer points to a regional identity. Then we move into the more dispersed official realm: the offices we visit, the places we telephone, addresses we write to with completed forms. They become places on our mental maps: we write to Swansea for our driving licence, to Liverpool for our passport, to Newcastle for our dole cheque, and so on. And finally, and running all the way through the referents noted here, there is the non-space of the media which is everywhere and nowhere and is rich in meanings. In the media realm we participate in imagined communities. And these are imagined as bounded.

The experience of memory

The sets of relationships that constitute identity are lodged within time, they have extension over time, they persist and they decay, and they do both at varying rates. The residues are lodged in memory. Private memory is the material of biography. The sets of ideas that will be affirmed by a group, the memories acknowledged by a collectivity, will be history; either the informal folk history of the group in question, or more formally as official history, or more scholarly as the history produced by historians. In other words, we can remember in a variety of ways. It is an active sphere. Samuel notes:[30]

> Memory, according to the ancient Greeks, was the precondition of human thought … memory, so far from being merely a passive receptacle or storage system, an image bank of the past, is rather an active, shaping force; that it is dynamic – what it contrives symptomatically to forget is as important as what it remembers – and that it is dialectically related to historical thought, rather than being some kind of negative other to it.

The notion of memory points to the ways in which practical activities deposit residues (individual and collective) in memory. It is a basis for ideas of continuity, a store of experience and knowledge to inform future activity. It is a sphere of reflective self-understanding. It is also a fluid sphere, liable to change/adulteration in the light of new events, or merely via the passage of time. We can distinguish personal memory, the stuff of autobiography and memoir, and collective memory, that is a contested public sphere. In the sphere of memory we make narratives that record, report and constitute our sense of our selves, our identity. Patrick Wright offers a wonderfully detailed report on the depth of the experience of change that is available in mundane routine activity:[31]

> What is to be done about Dalston Junction? Successive governments have pondered this question. Their advisors take one look and quickly propose a

> road-widening scheme or, better still, a really ambitious new motorway which will obliterate the whole area ... Dalston Lane extends east from Dalston Junction, and we need only follow it a few hundred yards up to the traffic lights at the next busy junction – a tangle of dishonoured roads still sometimes called Lebon's Corner in memory of a trader who has long since disappeared ... The south side of Dalston Lane starts with an elegant stretch of ornamented Victorian brick-work, which is all that remains of the recently demolished Dalston Junction railway station. It then passes a tawdry amusement arcade, a few shops and the New Four Aces Club ... After a derelict site and an ailing public library, the street consists of two continuous blocks of run-down Victorian shops ... The north side ... is slightly more varied. There are some shops with offices above them and an old pub, once known as the Railway Tavern but now a dingy betting shop with a satellite dish at the back. There is the notorious Dalston police station ... Then comes a terrace of stuccoed Victorian houses ... Beyond this residential terrace, there's a nondescript factory, a large and surprising Georgian house used as work-shop space by the Free Form Arts Trust, and, finally, a second Victorian ruin to match the shattered railway station ... The old vicarage of St Bartholomew's may be derelict, but it can still be said to command the north side of Lebon's Corner ... So this Gothic hulk stands there: a huge pigeon roost, a poster stand, a terrible warning of the destiny that awaits listed buildings in Hackney.

We inhabit cultural little traditions (and these change). Hoggart notes that the UK is still drenched in the marks of social class:[32]

> You may meet the sense of class every day and almost everywhere .. It is to be met in railway stations, supermarkets, work-places, among those taxi-drivers who have not yet decided to call all their fares 'guv' or 'mate' ... Yet change there is, and the mistaken blanket-belief that change has already come about blinds its adherents to its slowness and the resistances to it; and above all to the actual nature of the changes which are in fact in train ... Status, which seemed to have replaced class, has then changed back into a new form of class. The gates have closed again against openness.

It is clear that we dwell within cultural great traditions (and these change). However, as with class structures change is slow. Finally, a key use of the past, both informal and official, is when we invoke the past to explain and judge the present. Overall, we affirm ideas of progress (identity is thus lodged in devel-oping time). In the context, we can note a tension between the details of personal or community memory, and the generalized demands of mass culture. Many have argued that history in this sense is a zone of conflict between the powerful and their subordinates. It is a battle for historical memory.[33]

Identity and political-cultural identity

Identity is not given, it does not express an essence, it is carried in language and made and remade in routine social interaction. The learning/relearning in the private sphere of the locale will be direct, one will learn how to fill in the blanks in the schedule of questions as one moves through one's routines. The learning in the public realm will be a matter of the presentation of private self in public (one might say that learning one's way around the local neighbourhood is not the same as learning how to salute a flag). The contingency of social relationships and the subtle nature of language entails that identity is a fluid, shifting construct. It is clear that identity is contested. In the domestic private sphere the contested nature of identity will be a product of the power relationships between persons, in the family, locality and community. In the public realm the contested nature of identity will be a product of the power relationships between groups.

A political-cultural identity expresses the ways in which locate ourselves within those ordered communities within which we make our lives. It is the way we understand ourselves as members of political communities. It is the way in which private identity is presented within the public world. The ways in which agents construe their relationship to extant political-cultural structures will affect directly the extent to which such agents take themselves to be able to take control of their own lives. Such a political-cultural identity could take a series of broad forms: person-centred (how an individual construes their relationship to the community they inhabit); group-centred (how persons lodge themselves in groups and thereafter how a grouping of persons construes their relationship to other groups within the community); collectivity-centred (persons in formal organizations); and nation-centred (how persons ordered as imagined communities thereafter construe themselves in relation to others). In the sphere of political-cultural identity the contestedness of identity is centrally important. It is here that we can begin to speak of delimited-formal and pervasive-informal ideologies; great and little traditions; official ideologies; and the extent of the development of the public sphere (an institutionally ordered sphere within which rational debate might be undertaken). The ways in which individuals or groups construe their relationships to extant political-cultural structures will affect the extent to which such agents take themselves to be able to react creatively to structural change.

Power/authority I: construing the political community

A political-cultural identity is a way of construing a relationship to an ordered political community. As with identity, it is not given, but rather it is learned and relearned in routine social practice. A political-cultural identity is not an alternative to identity-in-general, it is merely an aspect of that identity – it can be discussed, as before, in terms of locale, network and memory.

The immediate locale will provide the person with a series of sources of knowledge (practices and ideas) in respect of the political community: family; neighbourhood; local area; organizations; institutions; and media realm. All these discrete spheres will provide knowledge that will inform the relationship of person and wider social world.

In the family sphere there will be the ways in which the members of the family severally construe their relationships to those centres and agents of power that they recognize. This could include the approach and judgements (practice and ideas) adopted towards those formal agents of authority encountered in routine practice (the agents of the state) – deferential, accommodative, evasive, aggressive or whatever. Hoggart recalls this aspect of family life and notes the routine:[34]

> dislike of petty officials, that seam or band of employees with whom working-class people have to deal about a whole range of matters in their efforts to be in touch with, get justice from, understand the shadowy powers who really run society … Petty officials, often uniformed which makes them even more off-putting, are likely to seem at the worst traitors to their class, at best uniformed boobies, more royalist than the king …

In the neighbourhood sphere of the street, the pub and the supermarket the person will encounter routine practices and ideas in respect of the politics of local life – how to deal with neighbours within the local sphere, and incomers/visitors to the local sphere (gossip, opinion, complaint), and the publicly voiced common-sense in respect of formal politics (for example commentary on topical issues in the media). Here the person operates as an individual and as a member of a small local group. In the wider local area the person will encounter both the lowest tier of formal political institutions and organizations (local state and local parties) and the lowest tier of formal political groups and social movements. Here the person operates as an individual and as a member of a group. Thereafter, in the organization sphere the person will encounter sets of ideas that express the relationship of groups to the community – as a member of a trades union or professional group. Here the person operates as a member of a social group.

In the institutional sphere of the state machinery – welfare, taxes, registrations of births, deaths, marriages and so on – the person operates both as an individual, a member of a group and as a member of the community. Yet this sphere is the primary reserve of an elite, those with authority. The majority are merely subject to this power/authority. The symbols of class pervade the machineries of the state. Nairn has picked out the institutional/cultural role of the monarchy in both exemplifying and helping to sustain the UK class system:[35]

> What is at issue here is a focal point of British culture … We're concerned with Royalty, one of the obsessive strains in that culture; but this obsession has a remarkably direct structural link to the second universally acknowledged fixation … class. Monarchy may be relayed to a British mass audience through an endless strip-cartoon of jokes, scandal, smut and Schmalz, from which 'the idea of Britain' seems remote. But in fact … the former depends logically on the latter and it is the latters' triangulation-points which compose the overall map: that is, the structure of authority which defines other speech, conduct and people … merely as those of class or region … 'manners' embody … the deeper structures of a society and State. All societies and States rely on such social customs and concrete verbal and body languages in reproducing themselves … The glamour of Royalty and the neurosis of 'class' are two sides of the single coin of British backwardness.

In the media realm the person participates vicariously in the issues of the day. In the adjacent realm of popular culture we find soap opera, celebrity, shock-horror stories and so on. In both, the agent operates passively – ideas are ready made, pre-packaged and popular opinion is confirmed (not debated). A similar vicarious participation is found in the sphere of popular consumption – the realms of supermarkets, shopping malls and life-style television programming – here there is also a clear political-ideological aspect: the ideas of market-place choice and political freedom have been collapsed. The linkage is routinely celebrated in media advertising. Jameson[36] elucidates the link between seeing the advertisements and buying into the ideology of market liberalism – the keys are the intertwined notions of choice, freedom and market. Jameson points out that the notion of the market is a political resource that serves the interest of the ruling groups. The pure market never did exist, so debate in respect of it is not debate about real social processes, rather Jameson tags the idea as being a crucial area of 'ideological struggle'.

The realm of media-carried consumerism ensures that the notion of the market is taken into common thought as both a given (natural) system and a realm of freedom. In this realm of market freedom, so the ideological story goes, the naturally given consumption desires of humankind find their expression in a fashion where, as Milo Minderbinder puts it, 'everybody wins'.[37] The ideological confection presents a core image and polices its boundaries. As Bauman says:[38]

> [the] strength of the consumer-based social system, its remarkable capacity to command support or at least to incapacitate dissent, is solidly grounded in its

> success in denigrating, marginalizing or rendering invisible all alternatives to itself except basic bureaucratic domination ... Indeed, as all the traditional demands for personal autonomy have been absorbed by the consumer market and translated into its own language of commodities, the pressure potential left in such demands tends to become another source of vitality for consumerism and its centrality in individual life.

These spheres of practice and ideas have been presented as a simple list of elements, however they are integrated. They find coherent expression in little tradition. Moving beyond this notion, and anticipating latter comments, the formal institutional aspect of political-cultural identity associated with these spheres has a cultural expression as 'great tradition' and/or 'official ideology' (the realm of state institutions and their several impacts upon persons). And most generally, and resonating with all these levels, we have the realm of media-carried political-cultural identity (the mass media of newspapers, advertising and television and their routine overlaps with consumerism, which ties them back into fundamental political-economic structures). In other words, political-cultural identity can be taken to be a coherent package, an integrated whole, which can be discovered in practice and ideas at micro and macro levels. It also finds expression in those networks that move beyond locale.

In terms of political-cultural identity we invest our immediate attention in the practice and ideas of locale that in turn are the basis for networks that move outside and beyond. We interact with distant others as individuals (family, holidays), members of organizations (commercial and professional links), and members of collectivities (trade links, international organizations). As we move along these networks our knowledge becomes attenuated – it loses detail – and at the extreme our knowledge of distant strangers falls away into stereotypes. It might be that we learn, it might be that we merely carry our prejudices with us, a distinction marked in the terms 'traveller' and 'tourist'.

A distinction can be drawn between public and private memory. In private memory we can distinguish between the unselfconscious understandings of particularity – the celebration of familiar patterns of life as unproblematic – and the reflexively self-conscious understandings of individuality – the celebration of the historical uniqueness of received and changing sets of circumstances.[39] And in the public sphere we have the realm of the national past – the agreed official common-sensical version of who we are, where we came from – a subtle construct, suffused with the demands of established hierarchies of power.[40]

The business of memory is linked to the matter of change. We recall change in a series of spheres: (i) the mundane details of family change, such as lines of relationship and patterns of descent (some formalized for property reasons); (ii) the ways in which the area/community in which we live changes over time, with new people and buildings; (iii) we note that cultural traditions shift and change, new ideas run through our lives; (iv) in all this the past is invoked to explain and judge the present; and (v) we habitually affirm ideas of progress, so identity is lodged in developing time. Once more we can note a tension between personal or community memory and tradition, and the generalized demands of official memory or mass culture. The business of memory involves both remembering and forgetting – it is not a passive exercise, rather it is active and suffused with conflict. History, in this sense, is a zone of conflict between the powerful and their subordinates.[41]

Patrick Wright argues that we can speak of a 'national past' – the set of stories that we take for granted – and that it is in these stories that we find a link between the particularism of our locale and that of the wider nation. Wright explains that:[42]

> Everyday life is the historically conditioned framework in which the imperatives of natural sustenance (eating, sleeping ...) come to be socially determined: it is in the intersubjectivity of everyday life that human self-reproduction is wedded to the wider processes of social reproduction.

In the routines of everyday life agents explain their world/selves via plausible tales, stories, narratives or histories. It is a part of the business of making sense. It is embedded in routine, the sphere of the taken-for-granted. It is also strongly implicated in the response of groups and individuals to change. Wright argues that the national past is a story told to inhabitants to render things meaningful in the context of the flux and disenchantment of the modern world. In the UK nostalgia plays an important role and Wright offers the examples of the model of the old-established family, with its below-stairs retainers, the use of craft models of manufacture, the invocation of old character types. All of these act to structure the present, as the inheritor and present expression of the national past. Yet Wright continues:[43]

> In the end, however, we come back to history in a more familiar sense, for the national past is formed within the historical experience of its particular nation-state. Among the factors which have influenced the definition of Britain's national past, therefore, are the recent experience of economic and imperial decline, the persistence of imperialist forms of self-understanding, early depopulation of the countryside, the continuing tension between the 'nations' of Britain (Wales, Scotland and, most obviously, Ireland), the continued existence of the Crown and so much related residual ceremony, the extensive and 'planned' demolition and redevelopment of settled communities which has occurred since the Second World War, and the still living memory of a righteous war that 'we' won.

The political issue for Wright lies in the question of whether or not:[44]

> … everyday historical consciousness might be detached from its present articulation in the dominant symbolism of nation and drawn into different expressions of cultural and historical identity.

Power/authority II: learning the political-cultural community

Political socialization takes place in a series of locations. It begins in the family. We can point to experiences in childhood with figures from outside the immediate domestic sphere – the neighbourhood, or local area – to whom parents or family members defer, or present requests, or display fear. Hoggart noted this, with visitors to the family home characterized as sympathetic, problematic, helpful, dangerous and so on. The experienced make these judgements quickly, but for the new recruit the realization that the social world is suffused with myriad patterns of power relationships is a long drawn out process of learning. It takes place in the domestic sphere, the immediate neighbourhood, factory or office and the wider local area. Thereafter we encounter the ideas carried by the machineries of the state – nation, patriotism and nationalism. A trio of ideas illuminate these matters: journeys, auratic sites and remembered war.

Anderson details the internal logic of journeys as constitutive of community (and their place in the growth of modern nationalism):[45]

> For our purposes here, the modal journey is the pilgrimage. It is not simply that in the minds of Christians, Muslims or Hindus the cities of Rome, Mecca or Benares were the centres of sacred geographies, but that their centrality was experienced and 'realized' (in the stagecraft sense) by the constant flow of pilgrims moving towards them from remote and *otherwise unrelated* localities … There was, to be sure, always a double aspect to the choreography of the great religious pilgrimages: a vast horde of illiterate vernacular-speakers provided the dense, physical reality of the ceremonial passage; while a small segment of literate bilingual adepts drawn from each vernacular community performed the unifying rites, interpreting to their respective followings the meaning of their collective motion.

Continuing, Anderson looks at the reactions of European elites to the growing extent and organization of empire and diagnoses a drive to independence conceived in terms of nationhood and sparked by the common experience of colonial functionaries, a new form of pilgrimage:[46]

> Though the religious pilgrimages are probably the most touching and grandiose journeys of the imagination, they had, and have, more modest and limited secular counterparts … the most important were the differing passages created by the rise

> of absolutizing monarchies ... The inner thrust of absolutism was to create a unified apparatus of power ... Unification meant internal interchangeability of men and documents ... Absolutist functionaries thus undertook journeys ...

And:[47]

> In principle, the extra-European expansion of the great kingdoms of early modern Europe should have simply extended the above model in the development of grand, transcontinental bureaucracies. But, in fact, this did not happen. The instrumental rationality of the absolutist apparatus ... operated only fitfully ... The pattern is plain in the Americas ... the pilgrimages of creole functionaries were not merely vertically barred ... lateral movement was as cramped ... Yet on this cramped pilgrimage he found travelling companions, who came to sense that their fellowship was based not only on that pilgrimage's particular stretch, but on the shared fatality of trans-Atlantic birth.

On the basis of the images of community thereby lodged in place, we participate in the 'national past'. Wright mentions two particular styles the national past adopts: the use of auratic sites[48] where history is taken as most particularly present; and remembered war which is taken as a sphere of non-routine when actions made a difference. Wright points out that there is a clear political aspect to all this as the group to whom we are routinely enjoined to belong/submit is the nation. We are offered a national past, an official history. The nation replaces other more local communities and it re-enchants the world. A story is woven which tells us who we are, where we came from and where we are going. The other matter of particular relevance is 'heritage', the authorized auratic residue of the past. Wright notes that it has expanded in recent years:[49]

> [and] now includes the local scene alongside the capital city, the old factory alongside the municipal art gallery, the urban tenement or terrace alongside the country house, the vernacular alongside the stately and academically sanctioned.

The cultural collectivity is learned. It is a matter of grasping the understandings embodied in institutions in the form of official ideologies (nationalisms), institutional truths and organizational truths.[50] It is also a matter of grasping the understandings expressed in routine practice: little traditions;[51] the ideology in everyday life;[52] and the everyday nationalism.[53] These themes are deeply lodged, and readily available and routinely repeated.

Linda Colley details the process of the invention of Britain, a variant on the theme of journeys. Colley details the elite's reinvention of itself in the wake of the defeat in the American War of Independence. The material vehicle of this reinvention was war and colonial expansion and the cultural vehicle was a celebration of heroic service of country. Colley speaks of the invention of

Britain by those who were heroes in their own epics. The UK elite constructed itself as a 'service elite' thereby creating a role and assimilating the future of the country to its own wellbeing. Colley comments:[54]

> the British elite's ability in the aftermath of the American war to associate itself with patriotism and with the nation in a new and self-conscious fashion proved invaluable ... The landed establishment's willingness and ability to change [was crucial] ... especially in the direction of greater Britishness ... By becoming a more unitary elite ... top Britons not only buttressed and consolidated their own social and political primacy, they also helped influence what Britishness was all about. Public schools, fox-hunting, a cult of military heroism and of a particular brand of 'manliness', the belief that stately homes are part of the nations heritage, a love of uniforms: all of these characteristic components of British life, which still remain powerful today, first became prominent under patrician auspices in the half century after the American war.

Power/authority III: acting in relation to the political community

The ways in which individuals or groups construe their relationship to extant political-cultural structures will effect the extent to which such agents take themselves to be able to creatively react to change. And the ways in which such agents understand their power will feed into their routine practical actions. There is scope for much conflict and confusion. Frank Parkin has identified a series of subaltern responses to the class system: deferential, aspirational, accommodative and oppositional. The opening two strategies derive from the dominant value system:[55]

> Deferential interpretations of the reward and status hierarchy stem from acceptance of the dominant value system by members of the subordinate class ... it tends to be bound up with a view of the social order as an organic entity in which each individual has a proper part to play, however humble ... Equally consistent with such acceptance is a view of the reward structure which emphasises the opportunities for self-advancement.

However, the resources of the subordinate class can also be used to construct an accommodation with the received hierarchy:[56]

> The subordinate value system tends to promote a version of the social order neither in terms of an open opportunity structure nor as an organic unity; rather, strong emphasis is given to social divisions and social conflict, as embodied in the conceptual categories of 'them' and 'us'.

Finally, it is only when an appreciation of difference turns into articulated action that one can speak of an oppositional meaning system, and this is the preserve of the politically active:[57]

> it presents men with a new vocabulary and a new set of concepts which permit a different translation of the meaning of inequality from that encouraged by the conventional vocabulary of society.

After Winch, we can say that agents can only act in the light of the understandings available to them. Views of the political community will be shaped by social context – by class, gender, ethnicity, family (or personal) circumstances. Our starting points for action will be varied – the starting points themselves will have multiple aspects (few people act in the light of some master-status). And all confront the givens of established social hierarchies and the machineries of the state – understanding entails action, but such action can take diverse forms, active and passive.

Change: threat or opportunity?

A wealth of material has been presented in recent years which suggests that identity is neither simple nor given, but instead is an elaborate social construction, made and remade in routine social practice. The identity of an individual person, or group or collectivity or community is fluid. It can shift and change. As circumstances and ways of grasping these circumstances change, so too does identity. One might think of identity as 'layered'. Some aspects are close to home and important (where the psychological costs of change are high). Others are more remote. Some are distant. The mix of elements comprising a personal identity can change over a lifetime. The same is true of groups. In this perspective, two familiar ways of speaking about political identity – class and nation – are misleading as neither class[58] nor nation are master statuses, they are merely layers of more complex identities. Once again, it can be argued that the best way to grasp identity is to look to the detail of the relevant constitutive, changing social processes.[59]

The end of the cold war in 1989/91 radically undermined received patterns of thinking. As the ideas of block-time fell away a new structural pattern, long in the making, was revealed. It comprised the development of an internationalized global system, a pronounced regional pattern and the project of the European Union. The confusions generated were numerous – end of history, clash of civilizations and globalization (for example). Amongst the British political elite there was alarm. The future looked European and Europe was a threat. The long-established strategy of dilute and delay was reaffirmed and redeployed. Tom Nairn[60] speaks of a 'late or terminal Britishness'. The end of the cold war turned out to be the start of an interregnum – 1989/91–2003 – a period of broad unease and confusion. In the UK (and maybe elsewhere in what had been 'the West') there was no political master narrative – Prime Minister John Major gave way to Tony Blair and a rolling wave of rhetoric ('spin') entranced the population but little happened. The interregnum came to a decisive end with the Iraq war. The British political elite opted decisively for America, against Europe and questions about the identity and trajectory of the polity, issues that had been managed, played down or deferred until some unspecified future date, were now out in the open.

Notes

1 The material here is derived from P. W. Preston 1997, *Political-Cultural Identity: Citizen and Nation in a Global Era*, London, Sage (see in particular, chapters 2, 3 and 4).
2 P. W. Preston 1996, *Development Theory*, Oxford, Blackwell.
3 After Laffan *et al.* we might speak of 'deep community'. See B. Laffan, R. O'Donnell and M. Smith 2000, *Europe's Experimental Union: Rethinking Integration*, London, Routledge.
4 Reading and reacting to enfolding structures will be partly *necessary* (after Peter Winch 1958 (2nd edn 1990), *The Idea of a Social Science and its Relation to Philosophy*, London, Routledge) – when our ideas change so has our world, it is not negotiable) – and part *contingent* – we can read and react and agree/affirm or disagree/resist (see A. MacIntyre 1962, 'A Mistake about Causality in Social Science' in P. Laslett and W. G. Runciman (eds), *Philosophy, Politics and Society, Series 2*, Oxford, Blackwell.
5 P. W. Preston 1981, *Theories of Development*, London, Routledge.
6 Z. Bauman 1988, *Legislators and Interpreters*, Cambridge, Polity.
7 Some technical (failings in the key nineteenth-century social science of political-economy in respect of market behaviour), some political (anxiety about the alleged socialist implications of political-economy) and some adventitious (resentments at claims to pre-eminence made by August Compte).
8 The argument is made in fuller terms in Preston 1997.
9 The arguments in this text do not derive from liberalism, and in what follows an individual is understood as the intersection of a series of social relationships that are reflexively monitored and given coherence in memory.
10 The force of 'ethnographic' is clear; as regards 'biographic' I have in mind first a simple strategy to order enquiry – the notion of identity is unpacked around the perspective of an ideal-typical individual, someone with a biography. The preference for detail of ethnography is thereby reinforced, and this links to a second thought: reflexivity is a necessary condition of scholarship – in respect of identity the background of the author is relevant. C. W. Mills (1970, *The Sociological*

Imagination, Harmondsworth, Penguin) pointed to intersection of private concerns and public issues as the spark of the scholarly imagination. Scholarship thus has an idiosyncratic element – it is written by people – and the analysts' interests/experience are part of the business of enquiry.

11 Hilary Mantel 2003, *Giving Up the Ghost: A Memoir*, London, Fourth Estate, pp. 70–1. I should add that I use lots of long quotes in this chapter – they illustrate the richness of identity – something with which we are all intimately familiar.

12 R. Hoggart 1989, *A Local Habitation (Life and Times Vol. 1: 1918–40)*, Oxford University Press, pp. 38–9.

13 Hoggart's work can be traced back – along with that of Raymond Williams or E. P. Thompson – to the work of F. R. Leavis. See R. Colls 2002, *Identities of England*, Oxford University Press, pp. 358–66.

14 Laurie Lee 1962, *Cider with Rosie*, Harmondsworth, Penguin, pp. 9–10.

15 Mantel 2003, pp. 127–9.

16 Lorna Sage 2001, *Bad Blood*, London, Fourth Estate, p. 21.

17 Hoggart 1989, pp. 124–6.

18 R. Hoggart 1995, *The Way We Live Now*, London, Chatto, pp. 229–300.

19 There is a large urban planning and architectural commentary on the city, the concerns of which are with design, patterns of life and social welfare. See Colls 2002, chapter 20.

20 P. Wright 1993, *Journey through the Ruins*, London, Flamingo, p. 13.

21 K. Waterhouse 1994, *City Lights: A Street Life*, London, Sceptre, pp. 125–8.

22 Sage 2001. p. 236.

23 Sage 2001. p. 270.

24 Hoggart 1989, pp. 79.

25 See P. W. Preston 1994, *Europe, Democracy and the Dissolution of Britain*, Aldershot, Dartmouth, p. 40 *et seq*.

26 Again a wealth of material – one thinks in particular of the work of Stuart Hall or Tom Nairn.

27 See for example, A. MacIntyre 1985, *After Virtue: A Study in Moral Theory*, London, Duckworth, or more broadly the political science literature which looks at policy communities, epistemic communities, policy networks and the like.

28 On Europe's memory, see Tony Judt 2002, 'The Past is Another Country: Myth and Memory in Post-war Europe' in J. W. Muller (ed.) *Memory and Power in Post War Europe*, Cambridge University Press.

29 Hoggart 1989, pp. 156–7.

30 Raphael Samuel 1994, *Theatres of Memory Vol. 1: Past and Present in Contemporary Culture*, London, Verso, pp. vii–x.

31 Wright 1993, pp. 22–4.

32 Hoggart 1995, pp. 199–200.

33 The potential conflicts are examined in Julian Barnes 1998, *England, England*, London, Picador, the tale of a property entrepreneur remaking England as a theme park on the Isle of White. The copy becomes the preferred version – a 'simulacra'. See also Umberto Eco 1987, *Travels in Hyper-reality*, London, Picador chapter 1.

34 Hoggart 1989, p. 127.
35 Tom Nairn 1988, *The Enchanted Glass*, London, Radius, pp. 61–2.
36 F. Jameson 1991, *Postmodernism: Or The Cultural Logic of Late Capitalism*, London, Verso.
37 J. Heller 1964, *Catch 22*, London, Corgi.
38 Bauman 1988, pp. 93–4.
39 From Patrick Wright (1985, *On Living in an Old Country*, London, Verso) – who takes it from Hannah Arendt. I think she is getting at a distinction between complacency and adulthood.
40 Again, Wright 1985.
41 In regard to Europe, see Judt 2002. For a novel see John LeCarre 1969, *A Small Town in Germany*, London, Pan.
42 Wright 1985, pp. 6–7.
43 Wright 1985, p. 25.
44 Wright 1985, p. 26.
45 B. Anderson 1983, *Imagined Communities*, London, Verso, pp. 55–6.
46 Anderson 1983, p. 56.
47 Anderson 1983, pp. 58–9.
48 See Peter Ackroyd (2000, *London The Biography*, London, Chatto) in which the entire city is read in auratic terms. The book was recycled on television in 2004.
49 Wright 1985, p. 25. In Japan they have the notion of 'living national treasures' and in Britain there are also popular national treasures, from Cliff Richard to Fred Dibnah.
50 Anderson (1983) discusses Britishness as an'official ideology'.
51 J. C. Scott 1985, *Weapons of the Weak*, Yale University Press.
52 Chua Beng Huat 1995, *Communitarian Ideology and Democracy in Singapore*, London, Routledge.
53 M. Billig 1995, *Banal Nationalism*, London, Sage.
54 L. Colley 1992, *Britons: Forging the Nation 1707–1837*, Yale University Press, p. 193.
55 F. Parkin 1972, *Class Inequality and Political Order*, London, Paladin, p. 85. See also D. Canadine 1998, *Class in Britain*, Harmondsworth, Penguin.
56 Parkin 1972, p. 88.
57 Parkin 1972, p. 90.
58 On the fluidity of class identification, see Canadine 1998.
59 On changes, see Preston 1997.
60 T. Nairn 2002, *Pariah*, London, Verso.

3 A development history

The sense that we have of ourselves, and our communities, flows from an elaborate creative exchange with the circumstances that enfold our lives. We inhabit sets of structured relationships. They are constituted in routine social practice. They are understood in the familiar vocabularies of ordinary life. There is no fixed reality, no given set of circumstances or asocial ground that can secure our social world. Nor is there any Archimedean point from which we might produce a definitive overview. We discover 'who we are' (we can articulate an answer) in retrospect.[1] All this is true of our routines. It is also true of the social scientific theories we use.[2] However, in contrast to the comfortable realms of common practice and common sense, social science must be sceptical, reflexive and engaged; that is, critical.

This short chapter sketches a development history of the UK.[3] It contradicts familiar ways of thinking about the history of the country. A development history points to the trajectory of a polity as it shifts down through time. It identifies the key structural patterns that impinged upon the form-of-life within the polity. It identifies the key agents within the polity and asks how they read and reacted to change, formulating new projects in order to move forward. The linkages of the polity with the wider regional/global systems will shift and change, as its importance and influence waxes and wanes. The territory of the polity will also change, expanding and contracting as events unfold. A development history is anti-nationalist. It assumes that change is given, forms-of-life and patterns of understanding are contingent, shifting and transitory and that periods of relative stability are interspersed with episodes of change (phases/breaks). When change does happen it can be rapid – a reconfiguration of structural patterns and embedded forms-of-life – there will then be newly empowered groups running novel projects.[4]

Trajectories: reading the past, characterizing the present

The official national past[5] of England – the 'Anglo-British' tale, the product of the historical subsumption of England in Britain - identifies the smooth linear process of the evolution of a coherent bounded community. The English emerge from the mists of time fully made and thereafter they make their way down to the present day. This is nationalist myth. In contrast a developmental history reveals a series of discontinuities – shifting economic patterns, changing social arrangements and fluid politics. The (changing) polity evidences a discontinuous sequence of locations within wider structural patterns – there is no single 'England', rather, as Norman Davies[6] points out, there has been a series of versions. Two ideas are important: territory and polity. The geographic territory of 'the Isles' has been home to numerous polities and they have had shifting borders. These polities have occupied some or all of this space and some have occupied space on the mainland. The present arrangement of territory/polities in the Isles is quite recent; the last major change was the secession of the Irish Free State in 1922.[7]

A development history identifies a series of stable phases. In each a particular territory is occupied. In each a particular political-economic, social-institutional and cultural pattern develops. In each a particular political-cultural identity develops amongst the elite (and is disseminated amongst the masses, intersecting with local little traditions). The identity is specific to the phase - a way in which the social world can be understood, ordered and maybe developed. The shift from one phase to the next will involve rapid complex change: (i) typically involving pervasive change in all sectors of the economy, society and culture; (ii) change whose end-point could only roughly be grasped; (iii) change ordered by the extant state-regime in terms of a distinct political-cultural project; and (iv) change which would be deeply unsettling for all concerned. A new phase has a new identity, and this process runs down through time. The present can be understood in terms of both the contingent accumulation of the social-institutional and cultural machineries put in place in response to changing structural circumstances (the identifiable residues of the past) and the pattern of contemporary creative agent responses to unfolding structural demands (the ways in which agents presently read and react to enfolding structural circumstances).

The recovery of the patterns of understanding characteristic of one phase is a matter of historical research, or a sort of historical ethnography. However, contemporary enquiry can only grasp the past in the terms available today (what else is there?), so hermeneutic understanding must of necessity be a specific endeavour. The transmission of elements of the identity of one phase to the next is both likely and contingent – contemporary research can uncover continuities and residues (there must be different ways in which the past of one phase can be transmitted into the future of a new phase), but once again must do so in terms of its current intellectual framework. What we have is a way of reading the past. The available intellectual resources let us detail the sequence of English and British polities which have been present in the archipelago of the Isles: (i) an early phase which saw the project of feudal England established; (ii) the mercantile capitalist English revolution which fixed the liberal project of England in place; (iii) the period of overseas reorientation in the eighteenth and nineteenth centuries which gave us imperial Britain; (iv) the episode of European civil war, which effectively destroyed the outward directed project of Britain; (v) the period of cold war, occupation and recovery that saw a remnant Britain subsumed within a US-centred sphere; and (vi) the contemporary phase of uneasy accommodation to the demands of internationalization, regionalization and the European Union.

Absolutism – the England of pre-modern late feudalism

The early phase saw the invention of 'England'. It was not created overnight. The broad sweep of the history of the Isles involves a sequence of peoples, civilizations and polities: the Celts were succeed by Romans, who in turn were succeeded by immigrations and invasions from northern Europe – Jutes, Saxons and Angles. Offa was styled the first king of the English in the eighth century and there were then are more waves of incomers. Alfred made a further attempt to establish a kingdom of England in the tenth century.[8] It did not succeed, and a Danish interlude followed. The eleventh-century Norman Conquest (another group of northern Europeans, settled in France) drew the Isles into the orbit of feudal France. At this time, there was no English elite, or masses or even language;[9] Davies reports that from 1154 to 1326 the Isles were part of feudal France.

The idea of England finally emerged in the pre-modern late feudal world of the fifteenth and sixteenth centuries as a regime based in the southeast of the Isles established itself against competition from other powers both within the Isles and on the mainland. Davies[10] points to a series of factors. Firstly, the extant polity was separated from France in the course of the Hundred Years War of 1337–1453 (a long episode of intra-European elite manoeuvring). In the war native dynasties came to the fore (Stuarts and Tudors) and provided a rallying point around which a polity in the Isles could coalesce. Secondly, the Reformation, the dissolution of the monasteries and the domestic wars of religion enabled the absolutist Tudor state to be strengthened, which it justified as recovering a pre-papal England (a myth as the Roman Church had long been present). The nascent southern Isles polity was cut-off from the mainland as it faced conflict with Spain, and re-imagined itself as 'English'. The Tudor state began the construction of an overseas 'English empire'; Davies speaks of 'the Englished Isles' (1326–1603). The emergence of the English language was the final element in this process. The use of 'middle English' spread in the fifteenth and sixteenth centuries (partly as a result of the Reformation, which provided access to God in vernacular tongues), and the idea of 'England' was popularized in the work of Shakespeare. Thereafter came further conflicts. Davies speaks of religious wars and the political attempts to unify the three kingdoms of England, Scotland and Ireland. However, we can point to the beginning of a materialist history at this point as the Civil War of the 1640s saw an end to the absolute monarchy, gave power to the landowners and brought about the rise of the towns – the establishment of agrarian mercantile capitalism.

The English Revolution – the emergence of mercantile capitalism

The English Revolution of 1642–88 marked the historical achievement of a mercantile capitalism, whose economy was essentially pre-industrial agrarian, with a society of nominally free labourers all ordered politically via a newly powerful parliament.[11] Over the seventeenth and eighteenth centuries the English polity developed domestically as a progressive liberal commercial society,[12] and internationally as a trading empire, as links were built up with mainland Europe and colonies in North America.

In conventional terms the start of the shift to the modern world in the Isles can be dated from this period in the middle of the seventeenth century. The work of Hobbes and Locke can be read as the contemporary expression of the shift from absolutist feudalism to a bourgeois mercantile agrarian liberal society.[13] A contemporary line of more radical work, anticipators of later ideas of democracy, is available in the work of the Levellers and Diggers.[14] The shift to mercantile capitalism was grasped in terms of the historically available models of the Renaissance city-states and the Dutch Staadtholderate. The English liberal political project was essentially mercantile, socially hierarchical and politically oligarchic – ruled by an enlightened elite. Yet all these ideas predated the democratic republicanism associated with the French and American revolutions – the world of the enlightenment. It was, thinks Nairn,[15] an essentially backward looking revolution, a partial and incomplete shift to the modern world.

Overseas reorientation – the invention of Britain

The 1707 Act of Union created a unified state within the Isles. The Scottish enlightenment flourished; the industrial revolution began and people moved south. The American War of Independence (1776) and the revolution in France (1789) radically disturbed the English pattern of development: there was no access to either America or the mainland of Europe; there were wars with revolutionary American and France; and there was domestic agitation for democracy from the Chartists. The political elite respond by reinventing themselves and their polity. It was the period of the invention of Britain.

The elite reinvented themselves as the rulers of an outward-looking trading nation. They were very successful at home, and the country began to industrialize rapidly. The social shift to the towns began and an urban society formed. The demands of the Chartists were defeated by 1848. Overall, domestically, the basis of an industrial class-divided hierarchical society was constructed, and overseas a large maritime global empire began to be assembled; this new project was up and running by the mid-nineteenth century. At this time, notions of Britain came to embrace the ideas of England, Scotland and Wales – what had been separate countries became provinces and so began the construction of a multi-national Britain. The shift to an industrial economy generated new class patterns and new domestic regional patterns (with areas of industry in provincial England), but the key centre of power was the southeast. The project developed through the late nineteenth century. There was further industrial advance (based in part on a scientific revolution), further urbanization, some social reform and a partial top-down sponsored democratization. The system reached its apogee in the years before the Great War.

Over this period the UK was seen as a model by other countries in mainland Europe, the Americas and in East Asia, in particular Japan. Reconstructing matters from the perspective of the present day it can be argued that measured in terms of the expectations lodged within the classical European tradition (ideas of progress, analyzed social scientifically in terms of dynamics of change) the UK was developmentally more advanced than its contemporaries.[17]

The European civil war

The history of Europe in the twentieth century falls into two inter-related phases. It is not the rise and fall of the European state-communist experiment that is crucial,[18] rather the long drawn out European civil war of 1914–45[19] with its attendant period of division and occupation (1945–2003). It was this that formed the context within which the British state-regime had to operate.

The Great War of 1914–18 caused widespread disruption: casualties, war damage and political upheaval. A series of mainland European empires were destroyed (Hapsburg, Hohenzollern and Romanov, plus the neighbouring

Ottoman). There were revolutions, successful in the USSR. The Versailles Peace Treaty required Germany to accept war guilt and pay reparations, but was widely criticized. The League of Nations was established, but without the USA and with little more than moral persuasion to enforce its views. A problematical system of new nation-states was established in Central and Eastern Europe: nations were to be fitted into states that were to have equal sovereignty in the new League. This arrangement formed many minorities, thereby storing up problems. In the UK the episode of war generated significant change: a loss of nineteenth-century optimism; a loss of economic and financial power to the USA and other competitors (Germany, Japan and others in Europe); and a defensive withdrawal into an empire trading sphere, itself increasingly disturbed by the first signs of the rise of colonial nationalisms. Although the Great War was followed by a brief period of optimism, it did not last. A series of failures followed in the 1920s and 1930s: the economic depression, the 'collapse of democracy' and the rise of fascism.[20] The late 1920s saw severe economic problems, exacerbated by orthodox economic responses, and the Wall Street Crash of 1929 precipitated a global depression. The democratic countries seemed unable to find a solution. There was a political turn to authoritarian national solutions – the rise of fascism – which in Italy, Germany and Japan was successful (at least in the short term). When Roosevelt came to office in the US an effective response was made – the New Deal. The US economy did not fully recover, however, until defence production began, and in the UK too it was defence production that began to get the economy moving. Thus the unhappy period culminated in the Second World War. It was this that marked an end to the political-cultural project of imperial Britain, bringing about the loss of empire and the ceding of economic and political leadership to the USA. A new historical development phase began.

Occupation, cold war and recovery – the English-speaking peoples, to 1989/91

The war in Europe came to an end in May 1945 as the armies of the USSR, USA and Britain reached standstill lines agreed by their political masters. A series of wartime meetings, plus one held immediately after the fighting, were taken to have settled in broad outline Allied plans for the future. In the event the agreements rapidly broke down, the cold war developed and the division of Europe into two blocks was fixed. There are many debates about the cold war, with some arguing it was all the fault of the Soviets,[21] others claiming it was planned in Washington[22] and some suggesting that it was a convenient smokescreen to cover great power occupations.[23]

The period 1945–89/91 saw occupation, division and the loss of a global role for Europe (including its remaining overseas empires) as economic and political power moved from European capitals to Washington and Moscow. The division of Europe into East and West saw two different sorts of polities develop – liberal-democratic (roughly) and people's democratic (sort of). The blocks developed elaborate block-ideologies, where each reflected quite directly the views of the relevant block-leader: in the east, ideas of state-socialism, in the west, ideas of free markets and free societies – the elaborate ideological confection of 'the West'.[24]

The situation of Europe internationally was parlous. The pattern of empire holdings was untenable and pre-war trading relationships were disrupted. A full appreciation of the new circumstances took some time to achieve: wars in Indonesia (Dutch); withdrawal/partition in India (British); guerrilla conflict in Malaysia (British); wars in Indo-China (French); wars in Algeria (French); wars in Suez (British/French); wars also in Angola/Mozambique (Portugal); Congo (Belgium); and a host of smaller areas of trouble. However, the fate of the European empires was perfectly clear: they were finished. The ethic of empire was unsustainable, there was a series of local nationalist movements and the Europeans were unable to fund the military forces necessary to insist. The key to European economic recovery was the Marshall Plan, which made credits available to fund European purchases in the US. With this facility in place the economic recovery of Europe was rapid: in Germany, the Wirtshaftswunder; in France, Le Plan; and in Italy, la dolce vita. In time, on the mainland, the EEC proved successful. In the UK, as the empire was stripped away, economic, financial, military and cultural priority passed to the USA, and the elite acquiesced in a subordinate status within the American-sponsored liberal trading system.

In the years following the Second World War the British political elite pursued a political-cultural project that acknowledged its new situation, but did so in a minimalist fashion as it endeavoured to sustain its pre-war status. The political elite's residual 'great power' aspirations, expressed in Bevin's 'Churchill option', met their end in the 1956 debacle of the invasion of Suez. The episode marked the end of the project of imperial Britain. However, consolations were found. The political classes reconciled themselves to US pre-eminence. Britain was rethought, no longer an empire but merely a long established liberal-democracy amongst other similar countries (save for the British aspiration to priority in comparison with mainland countries). It was the period of the post-war settlement, the idea of 'the West'. The tale worked after a fashion until 1989/91.

The shock of the new – 1989/91–2003

The post-Second World War phase of the development history of the UK came to an end in 1989/91. The end of the cold war meant the end of the role of number one ally. The end of the rhetoric also revealed the existence of a tri-regional global system: North America, East Asia and Europe. The British political elite confronted the task of reading and reacting to the new situation, while the European Union became the central project around which politics revolved. The British elite endeavoured to adjust to the project of the EU (thus, symptomatically, Margaret Thatcher was sacked and John Major signed the Maastricht Treaty in 1992). However, overall the 1989/91–2003 period represents something of an interregnum[25] – one set of ideas about how the world was ordered was overthrown but there was no clear replacement in view. General debates took place – the end of history was announced, as was the clash of civilizations, each offering ways of continuing the predominance of the USA within the sphere of 'the West', but these were always implausible schemes. The slow rather directionless debates within the residual West were re-energised by the attacks on the USA of 11 September 2001 (9/11). The American government had a choice: read the attacks as criminal and invoke the law (UN, ICC, security measures, counter-terrorism and diplomacy), or read the attacks as war, and call in the military. The choice was made in a few days: the government of President G. W. Bush quickly confected[26] an enemy and attacked Afghanistan. It was a mix of revenge, demonstration and opportunism,[27] but in the wake of 9/11 the American government had wide international sympathy and support.

In the run up to the American 2002 mid-term elections, the long-running Iraq problem was moved to the centre of the Washington agenda. Whilst critics diagnosed electoral cynicism, the influential US neo-conservatives had wider ambitions. The government promulgated doctrines of unilateralism and preventive war,[28] and attacked Iraq in March 2003. The interregnum was ended; the debate about the post-cold-war world was now up and running. Once again, problems were generated for the British political classes: their disposition was to ally themselves with the American government, yet domestic popular opposition was strong, as was international opposition. While the British political elite rallied to the support of President Bush, in Europe the German and French political classes emphatically rejected such a line. Throughout Europe there was widespread popular opposition to the war, and as the Iraq crisis unfolded in the spring and summer of 2003, with active military operations giving way to the concerns of long-term occupation, the British political elite found itself in an unenviable position.

What next?

The circumstances that enfold the lives of agents shape political-cultural identities, through the mixing of public demands and private understandings and aspirations. These in turn inform action. The demands of circumstances can register with the agent in one of two ways: obligatory or voluntary. In the first the agent reacts as soon as the situation is understood; in the latter we grant that there are many situations in which recognition can be distanced from the agent. Space can be made to grant the claims, reject them or, in the case of structurally weak positions, to subvert them (to deploy the weapons of the weak). None the less, whatever the dynamic of the unfolding processes, it remains the case that changing circumstances generate changing understandings, requiring novel lines of action. In the case of the UK population two ways of reading current circumstances can be indicated: first, the nationalist tale of the smooth accumulation of the institutional mechanisms necessary to sustain a continuing deep identity within the modern world of nation-states ('a thousand years of history', etc.), a conventional (and deeply familiar) nationalist British history which finds an essential continuity down the centuries. Second is a development history which identifies a series of polities within the Isles, each a way of reading and reacting to enfolding circumstances. At the present time the familiar post-Second World War scheme of 'Britain' is under threat. The cold war context within which the project was elaborated had already changed – both internationally and domestically[30] – but the period 1989/91–2003 has made it redundant.

These upheavals mean that the location of the UK within the modern global system is now open to critical inspection in a way not available during periods of relative stability. The context within which any UK political elite must operate comprises the internationalized global system, the pattern of regions and the locally dominant project of the European Union. Similarly, just as the location of the UK is open to inspection, so are the goals the polity affirms. The post-Second World War preference for a subordinate linkage with the USA – recently emphatically re-affirmed by the British political elite – is not merely not the only option on offer, it is the more implausible one. As links with the mainland deepen – economic, social and cultural – the 'route to the future' for the UK population is the project of Europe – unclear, contested yet ever more present in the lives of the inhabitants of the community.

Notes

1 After Hannah Arendt (cited in Patrick Wright 1985, *On Living in an Old Country*, London, Verso) we can shift from 'particularity' to 'individuality' or, as Hilary Mantel has it, we can cease to be written by our families and younger selves and

take control of the copywright of our own lives. Hilary Mantel 2003, *Giving Up the Ghost*, London, Fourth Estate, p. 70.

2 This can be understood after the style of Jurgen Habermas. It can also be understood in a rather different way – see Maureen Whitebrook 2001, *Identity, Narrative and Politics,* London, Routledge. The realm of literature can help, she argues, in political life/analysis, as the business of telling a story is relevant: narrator, voice and point of view are all involved

3 The notion of a 'development history' or 'trajectory' is mine. It is derived from development theory, but the substantive history is taken from the work of social historians. I have taken liberties with their sophisticated and nuanced work in order to make the point clearly.

4 Alternatives stress continuity – see J. C. D. Clark 2003, *Our Shadowed Present: Modernism, Postmodernism and History*, London, Atlantic Books.

5 Wright 1985.

6 N. Davies 2000, *The Isles: A History*, London, Papermac; N. Davies 1997, *Europe: A History*, London, Pimlico. On English historiography see also E. Jones 2003, *The English Nation: The Great Myth*, Stroud, Sutton Publishing.

7 Davies 2000, p. xxxvii. See also K. Robbins 1998, *Great Britain: Identities, Institutions and the Idea of Britishness*, London, Longman, pp. 277–84.

8 Davies 2000, pp. 226–9.

9 Davies 2000.

10 Davies 2000.

11 I take the argument from Christopher Hill and the other British Marxist historians.

12 R. Porter 2000, *Enlightenment: Britain and the Creation of the Modern World*, London, Allen Lane.

13 C. B. Macpherson 1962, *The Political Theory of Possessive Individualism*, Oxford University Press.

14 H. Kaye 1984, *The British Marxist Historians*, Cambridge Polity. Christopher Hill was a key figure.

15 T. Nairn 1988, *The Enchanted Glass*, London, Radius. The point is debated – critics say Nairn has the model of the French revolution too much in mind. However, whilst one might say that the UK looked advanced for a long time, by the late nineteenth century that was no longer true, and today the structure of the polity does look outmoded.

16 See L. Colley 1992, *Britons: Forging the Nation 1707–1837*, Yale University Press.

17 Arno Mayer 1981, *The Persistence of the Old Regime*, New York, Croom Helm.

18 Shortly after the end of the cold war, Eric Hobsbawm published his analysis of 'the short twentieth century', covering the period 1917–89 (E. Hobsbawm 1994, *The Age of Extremes: The Short Twentieth Century 1914–1991*, London, Michael Joseph). In this scheme the episode of the historical presence of the Soviet Union frames the history of Europe – Great War, revolution, depression, Second World War, long boom followed finally by further uncertainty. The analysis is persuasive, yet there are (obviously) other ways to think about the twentieth century history of Europe. In a wider, global, context it is clear that the history of Europe does not revolve around the USSR.

19 Arno Mayer (1988, *Why Did the Heavens Not Darken: The 'Final Solution' in History*, New York, Pantheon) speaks of a 'general crisis': 'The elites and institutions of Europe's embattled old regime were locked in a death struggle with those of a defiant new order: in the economic sphere merchant and manufactural capitalism against corporate and organized industrial capitalism; in civil society prescriptive ruling classes against university trained strategic elites; in political society land-based notables and establishments against urban-based professional politicians; in cultural life the custodians of historicism against the champions of experimentation and modernism; and in science the guardians of established paradigms against the pioneers of the world's second great scientific and technological revolution ... It coincided with the loss of Europe's primacy in the world system and its retrenchment from overseas empire ... Europe's general crisis helped generate enormous foreign wars ... it fuelled the two world conflicts ... By a feedback process these monstrous wars intensified and accelerated the antecedent crisis' (pp. 3–4).

20 On all this see Mark Mazower 1998, *Dark Continent: Europe's Twentieth Century*, New York, Alfred Knopf.

21 J. L. Gaddis 1997, *We Now Know: Rethinking Cold War History*, Oxford University Press.

22 G. Kolko 1968, *The Politics of War*, New York, Vintage.

23 R. Aron 1973, *The Imperial Republic: The US and the World 1945–1973*, London, Weidenfeld.

24 The 'West' comes in varieties – see Davies 1997, pp. 22–5, 40. See also Clark 2003, chapter 7, who argues that the idea is American and mis-analyzes European history.

25 Tony Judt 2002, 'The Past is Another Country: Myth and Memory in Post-war Europe' in J. W. Muller (ed.) *Memory and Power in Post War Europe*, Cambridge University Press.

26 The elements included the Taliban government of Afghanistan, the Islamist ideology of al-Qaeda and the person of Osama bin Laden, rhetorically presented as a coherent hierarchical group based in Afghanistan.

27 See P. W. Preston 2002, '9/11: Making Enemies; Some Uncomfortable Lessons for Europe'. Paper presented to the *European Union in International Affairs Conference*, Australian National University, 3/4 July.

28 The UN and international law allows for pre-emptive attacks when it is clear that an enemy is about to attack. The notion of preventive war – attacking those whom you suspect might at some stage in the future be a threat – is new and clearly the notion radically undermines the ideas of the international system. In common parlance the term 'pre-emptive' tends to be used to mean 'preventive'.

29 D. Marquand 2003, *New Statesman,* March; also D. Marquand 2003, *New Statesman*, 24 November.

30 See for example Yasmin Alibai-Brown 2001, *Who Do We Think We Are: Imagining the New Britain*, Harmondsworth, Penguin. Or see Zadie Smith 2001, *White Teeth*, Harmondsworth, Penguin.

4 The project of 'Britain'[1]

The political-cultural project of 'Britain' took shape in the late eighteenth and early nineteenth centuries.[2] It was an elite construct, a response to difficulties in America, mainland Europe and the domestic arena. An oligarchic liberal regime engineered the creation of a global trading empire, the 'British world'.[3] It was part of an unfolding historical trajectory, inevitably contingent. A number of lines of argument can illuminate the construction of the project of Britain: political-economic, which points to the macro-structural occasion of the invention of Britain; social-institutional, which point to the institutions which carry established patterns of life (the ways in which the elite has read and reacted to the demands of change); and culture-critical, which points to the spheres of high and low culture whereby individuals and groups have reflexively understood themselves. These arguments are distinct and do different sorts of work, but they can be run together to produce a substantive statement[4] that details the historical occasion, unfolding dynamics and contemporary circumstances of the British polity. A final concern, informed by these lines of reflection, is the matter of political-cultural identity, the ways in which members of the British polity understand themselves as political agents – what is it to be British; what might it be to be English?

Structures, agents and projects

The shift to the modern world saw the formation of states that subsequently invented nations.[5] A polity would be legitimate in the eyes of its inhabitants if it were committed both to material advance and rule by those co-cultural with the ruled.[6] Nation-statehood can be read as a project. After Benedict Anderson[7] we could speak of imagined communities having anticipated futures. But the project is not fixed, it is contingent and it can change. Recently theorists have considered 'post-national' and 'post-state' patterns of political organization (matters pursued with respect to ideas of globalization and regionalization, in particular with the EU) however, for the moment, we can run with the familiar idea of a territorially bound nation-state.

As noted above, the modern UK polity emerged in phases.[8] Norman Davies[9] argues that an important separation from the mainland followed the Hundred Years War (1337–1453). The Tudor state asserted its power, feudalism was superseded with absolutism, England became the elite's political project, and an outward-looking English empire was created. In the seventeenth century, the English revolution established an agrarian mercantile capitalism in which parliamentary absolutism secured freedom for the propertied. A period of rapid progress followed, an 'English enlightenment'.[10] Thereafter, in the eighteenth century the system incorporated Scotland, expanded overseas, defeated subaltern republican democratic aspirations and invented Britain and British nationalism.[11] In the nineteenth century, industrial capitalism emerges, there was global expansion and by the end of the century the British polity had attained its familiar institutions, class groupings and patterns of routine political conflict and debate. The tale thereafter is a familiar one – the twentieth century saw wars, decline, occupation and the elaboration of familiar consolations.

The pre-modern world

Davies offers a characterization of the pre-modern world. He remarks that it is difficult to imagine the Middle Ages:[12]

> Tiny, isolated settlements existed in an overpowering wilderness of forest and heath, in a stillness where a church bell … could carry for miles … People's perceptions … lacked any strong sense of discrimination between … fact and fiction … The Medieval awareness of a time and space was radically different … Above all … people lived in a psychological environment of fear – man was feeble and God was great.

It carries over into identity. Davies points out that this period was pre-national:

> Medieval Europeans were conscious of belonging to their native village or town, and to a group possessing a local language … who acknowledged the same feudal lord … [who belonged] above all to the great corporation of Christendom.

The territory of the Isles was home to a series of peoples: first the indigenous Celts, and thereafter a series of migrants – Angles, Saxons and Norman French, to name but a few. All were pre-English. The population was low, a few millions. The Isles saw a series of polities, some of which were self-contained within the archipelago (the earlier ones), but most overlapping with the mainland. In the years before the invention of 'England' most of the territory of the Isles was part of the political system of the French kings.

Absolute monarchy

The political history of the Isles has been intertwined with the history of the mainland.[13] Against the claims of British nationalists to a separate reality, the archipelago has long been an integral part of European history.[14] Indeed, for most of that time, the territorial polity has encompassed lands on both sides of the narrow strait that separates the main island from the mainland.[15] However, the relationship with the polities of the mainland has been fluid. The Hundred Years War, read by Davies as elite intra-feudal manoeuvrings, produced a division between the Isles and mainland. The native dynasties of Tudors and Stewarts competed for power and England becames the political project of the southern elite. The break with Rome of the Reformation strengthened the Tudors, the Tudor state asserted itself and feudalism was superseded with absolutism; there were parliaments and lots of assemblies, but the monarchy held power. The language of English spread and a national myth was created which celebrated the Tudors and the English. This period was the starting point of an Anglo-centric perspective that systematically read the mainland out of the history of the Isles.[16]

Parliamentary absolutism

The shift to the modern world,[17] the change from one form-of-life to its historical successor, was a long drawn out process. The transformation of agrarian feudalism into mercantile capitalism required the political form of the sovereign state, which in turn needed an appropriately mobilized population, secured via the idea of nation.[18] The ideal of popular nationhood emerged in the eighteenth century. It offered a novel layer of identity, a defined grouping of similar people, a political community.[19] These changes in economy, society and polity were bound up with ideas of enlightenment, the promise of the experience of revolutionary America and France.

If this is the familiar model of the shift to the modern world, it is clear that the English polity began to change much earlier – before the ideas that animated the American and French experience found wide expression, before the patterns of life that carried these changes found wide expression, before the celebration of reason at the heart of the enlightenment provided models against which reformers might orient their programmes (indeed, the English experience provided a model for the French theorists of the enlightenment[20]). The precise nature of the changes in the Isles has been much debated (primarily by historians) and three strands of argument stand out: an Anglo-centric line, unpacking the serial particularities of the evolution of the English (later British) – kings, queens, wars, religions and lately the appearance of industrialism and liberal democracy; an historical-materialist line, focused on the detail of the experience of the people of the Isles; and a more politically engaged line of Marxist analysis.

The historical-materialist notion of an English revolution owes much to the work of a group of historians: Christopher Hill, Rodney Hilton, Edward Thompson and Eric Hobsbawm.[21] The shift to the modern world in England is read in class terms. The English revolution of 1640–88 secured the property rights and freedoms of the new agrarian mercantile bourgeoisie via a new parliamentary absolutism. The feudal system centred on an absolutist monarchy was displaced (the settlement of the Tudor period was revised), and a proto-democratic movement in the form of the Levellers, Diggers and so on was defeated. The 1688 political settlement successfully ensured freedom for the propertied, and the eighteenth century saw widespread economic, social and cultural progress.[22] The English polity became the model of a progressive society for subsequent enlightenment political thinkers.

In recent years, more orthodox historians have challenged these analyses,[24] but the broad argument seems secure. A rather different line of criticism is also available however, centring on the timing and character of UK capitalism. Tom Nairn argues that it is a mistake to read the early modern history of the UK in terms of the ideas of modernity, nation-state and democracy, the core ideas of contemporary political thinking. These ideas only entered political discourse after the development trajectory of the polity had been given preliminary expression and set on a definite course. The project of liberal England in the seventeenth century did not look forward to the modern world, it looked backwards to the city-state republicanism of Renaissance Venice, or the seventeenth-century Dutch Staadtholderate: oligarchic, hierarchical and mercantile.[25] It was the later eighteenth-century enlightenment in Scotland, France and Germany, together with widespread admiration of the achievements of liberal England,[26] that inspired the revolutions in America and France. It is their ideas of republican democracy and the necessity of progress that provide the crucial elements of European political discourse. In the nineteenth century the British elite reacted with sustained hostility[27] to the enlightenment ideals of democratic nation-statehood. The historical trajectory of the Isles produced a pre-democratic class state distinct from the absolute monarchies in power on the mainland and the later bourgeois nationalist states sought by the American and French revolutionaries. Nairn draws the political lesson: 'It is the distinctive political coordinates of the early-modern that provide a definite historical location explaining both the half-modern and the unmistakably archaic aspects of twentieth century Britain'.[28]

The invention of Britain

The eighteenth century saw the invention of Britain. The incorporation of Scotland in 1707 created a unified state and market within the island archipelago. There was rapid economic advance. The country was prosperous. It had links to the mainland and a collection of colonies in North America.[29] Roy Porter[30] celebrates the robust commercial vigour of the period, which generated significant scientific/cultural advance[31] and became a model for mainland thinkers. A complex series of changes unfolded over the course of the century.

Linda Colley[33] considers the slow emergence of Britain in terms of the contemporary dynamics of structures/agents. The 1707 Act of Union produced an influx of Scots – they came south to find their economic/political fortunes in the newly expanded domestic marketplace. The new political/economic unit was called 'Britain'. As Colley suggests, being 'British' was a good idea because it offered people a route to future prosperity. The archipelago was also a part of a wider Atlantic economic sphere. The American Revolution of 1776 was a significant economic and political loss to the UK – it meant the loss of colonies and large parts of North America were closed off to English traders. At the same time, as Colley points out, there was hostility towards the mainland – both religious conflicts (Catholicism/Protestantism) and the wars that unfolded downstream from the 1789 French Revolution. The conflicts with revolutionary Europe blocked access to the mainland. It was this set of structural problems that generated the political-cultural project of 'Britain'. Colley[34] presents the structure and discourse of 'Britain' as a project occasioned by this loss of North American colonies, the hostility to the mainland and the expectations of enhanced material wealth. The elite political classes re invented themselves, turning outwards. The integrated domestic market provided a platform for overseas trade and the accumulation of a global empire. After the trauma of the American rebellion the new empire was not English, but came to be inclusively 'British'. Throughout the nineteenth century the project flourished on the back of military success, economic innovation and political repression.

Over the period 1790–1820 the emergent British polity, with its received legitimating discourse of sovereign-in-parliament, was challenged by the modernity of the French enlightenment. The ideals of reason, progress and republican democracy (citizenship and equality) found an echo in the subaltern classes. The Chartist movement sought to democratize the nascent project of Britain, and they sustained their arguments until the 1848 'springtime of nations' when the political elite suppressed the movement.[35] Although the defeat of Chartism blocked domestic democratic reform, the experience of rapid industrialization (with its characteristic admixture of dislocation and advance), interstate warfare and the achievement of strategically significant military victories, presented the ruling oligarchy with a favourable context for regressive modernization.

The nature of British capitalism has been much debated. The Nairn/Anderson thesis asserts that a variant with quite particular characteristics took root. Critics have suggested the authors were over-impressed by the revolutionary upheavals associated with the mainland, France in particular, and thus mis-read the domestic scene, which was, when considered directly (rather than through the prism of a particular view of the French experience), economically, socially and politically vigorous.[36] The vigour is not in doubt; rather it is the nature of the social dynamic.[37] Nairn argues that Marx was right about the production dynamic of capitalism but was wrong about the British polity fitting that model, because it was essentially a mercantile capitalist system with industry as an extra.[38] It was a quite particular form of capitalism, essentially a pre-industrial capitalism driven by the accumulative dynamic of mercantilism. And it had a distinctive polity. Benedict Anderson[39] argues that as the UK shifted to a modern-type nation-state comparatively late the nationalism – Britishness – was always an official nationalism, constructed and imposed from above,[40] and to which the monarchy, represented as the symbolic head of 'the nation', was central. Nairn argues that with the defeat of the Chartists, notions of republican democracy dropped out of subaltern critical political discourse. The period saw the making of the English working class,[42] but it was a de-mobilized class, its concerns re-directed to trade union-based ameliorist welfare programmes (pursued in concert with a spread of middle-class social reformers). An oligarchic ruling elite based in the southeast of England meanwhile sustained its control of an outward-looking global trading system.[43]

A liberal-oligarchic system in place[44]

The nineteenth century saw the rapid development of industry, considerable growth in urban settlement, significant socio-political reform and the promotion of empire-nationalism. In the later years of the century Britain continued to act as the metropolitan core of a global trading network which had expanded into new territories in Latin America, East Asia, the Middle East and Africa. The project reached its peak in the years before the Great War, in terms of domestic industrial and scientific advances, the overseas empire and the domestic promulgation of a grandiose British nationalism,[45] a 'High

Imperialist variant of "Greatness"'.[46] Thereafter, however, the outward-looking southern bourgeoisie continued to prosper whilst the metropolitan unit in total – the UK – entered a period of long and continuing relative decline.

The English revolution had put into power a landed and commercial bourgeoisie. In responding to the multiple challenges of the late eighteenth and early nineteenth centuries the ruling groups fixed in place an outward-looking mercantilist capitalism, and secured their system and their control via a monarchy-centred British nationalism. While the modernist ideals of republicanism were broken along with the Chartists, however, the powerful new industrial sector took longer to absorb. There had been a strong provincial economy and society that had produced its own anti-centralizing, bourgeois radical social and political ideas,[47] but the southern trading bourgeoisie asserted their authority, and just as any possibility of republicanism had been suppressed, any possibility of an industry-centred modernity was also lost. Metropolitan London had become the key centre of power by the late nineteenth century and provincial Britain, with its great cities, was eclipsed. It was in this period of reform and ruling-class readjustment that the familiar shape of British political discourse was settled.

The presently familiar institutions and ideologies of the British polity – the Westminster model[48] – were fixed in place, as was the pervasive conservatism and the commitment to a broadly liberal ideal of market, society and polity. They have stayed in place more or less unchanged ever since, surviving two world wars, the ending of an empire system and absorption into the US sphere. In regulationist terms,[49] the British political class (a shifting fractional alliance) has successfully retained control of the domestic polity – and thus economic, social, political and cultural power – whilst responding, by and large successfully in class terms, to the successive reconfigurations of the international system.[50]

The slow pace of change

The European classical social theorists were concerned to elucidate the dynamics of complex change in order to inform the modernist project of the achievement of a rational democratic society. The analyses that they made were holistic, prospective and engaged. An element of the logical structure of these enquiries was an implied general theory that offered a narrative of the theory translated into practice by its putative agents, a tale about how things were and how they might be expected to work out over future years. There were expectations about the pace of change; expectations about the historical project of modernity. If events are measured against these time-scales then it is still *far earlier* than anyone might have anticipated.[51] The post-absolutist ancien regime, pre-bourgeois and pre-industrial, did not leave the historical stage smoothly, rather it resisted and fought back. Arno Mayer[52] argues that the Great War was precipitated by this refusal to leave the historical stage. The large monarchical empires – the Habsburgs, Hohenzollerns and Romanovs – continued their autocratic rule over much of the European mainland until their collapse in the chaos of war.[53] Thereafter it was not bourgeois democracy that came into its own, rather the ferocious reaction of European fascism. The whole period from 1914 through to 1945 revolved around the dissolution of Europe's old regimes. Mayer[54] characterizes the era as a 'general crisis'; complex strains precipitated a series of wars that devastated Europe.

The pace of general systemic change was slow. The British political-economic, social-institutional and cultural systems had looked comparatively advanced within the context of the extant capitalist societies throughout much of the nineteenth and early twentieth centuries. However, British pre-eminence was in process of eclipse from the Great War, and in the interwar period Britain withdrew into an empire-centred sphere; it proved to be an illusory respite from relentless pressures for change.

The years between the two wars

The early part of the short twentieth century[55] was a time of conflicts and defeats for Europeans. In the space of thirty years they were involved, sometimes directly, sometimes more peripherally, in a series of wars: the Great War (1914–18); the Chinese Civil War (1928–49); the Sino-Japanese War (1931/37–45[56]); the Second World War (1938/39–45[57]); and the Pacific War (1941–45). The European wars were viewed at the time as international, that is, between separate nation-states, but they might better be seen as an interlinked series, together constituting a European civil war (1914–45): Arno Mayer's 'general crisis'.[58] It was a chaotic period, one of widespread destruction and great loss of life.[59] The countries of Europe found themselves eclipsed as power shifted to Washington and Moscow,[60] and the sometime familiar claims to European civilization and pre-eminence were rendered unequivocally absurd.

In Europe the Great War had a series of impacts. The continent was no longer the dynamic centre of global capitalism. European self-understandings were radically disturbed,[61] political life was disordered[62] (through both domestic upheavals and new centres of international power). In the midst of this came revolution, depression and fascism. In Russia the old dynasty was overthrown amidst the chaos of war, causing a series of widening circles of reaction, including elite shock in other parts of Europe, Allied intervention against the Bolsheviks (involvement in the Russian civil war) and other attempted revolutions (in Germany, Hungary and Austria). The economic recovery of the 1920s did not last: the 1929 stock market crash[63] precipitated a global depression and the advanced economies saw dramatic falls in output and heavy unemployment. The policies adopted by governments to alleviate the problems – cutting spending, balancing budgets and erecting trade barriers – only made matters worse and contributed to the rise of diverse fascisms. The fascist movements themselves were hostile to the available pattern of modernity and deployed an alternate set of ideas: conservatism, romanticism, aesthetico-mysticism, militarism, racism, anti-Bolshevism and the celebration of will/action. As a political-cultural project fascism came in varieties: (a southern European Catholic clerico-fascism (Italy, Spain and Portugal), and a northern European irrationalist/ideological fascism (Germany), as well as fascist-style movements in other European countries (Ireland, Hungary, Romania and Britain).[64] While there were a series of reactions to the rise of fascism in other European countries, the memory of the chaos of the Great War inhibited direct responses: the French were uncertain and the British contrived a policy of appeasement.[65] In the event, the history of European fascism is brief. The core countries were Germany, Italy and Spain, and in each case the fascist movement came to power utilizing extra-parliamentary violence, and the local populations provided their first victims. The slide to war gathered pace and from 1938/39 disaster began to unfold.[66]

In Britain the direct experience of the Great War was traumatic. An early optimism was disappointed as the realities of industrialized warfare and military incompetence generated enormous casualties (read into public experience in terms of the trenches, myths of lions led by donkeys and ideas of needless sacrifice). An extensive domestic mobilization was organized, yet after the war political promises of material reward went largely unfulfilled (not enough 'homes fit for heroes'[67]). The tensions found expression in the 1926 General Strike. The failure of liberalism, the example of the success of the USSR[68] and the early success of fascist regimes are also noted.[69] There was political drift, economic advance (with unemployment concentrated in old industries) and a continuing social system of hierarchy and conservatism. The post-Great War period also saw a vigorous debate amongst intellectuals and policy analysts about the shape of the future: contributors included J. M. Keynes, E. H. Carr, Karl Manheim and William Beveridge.[70] A current of reaction was present in elite sympathy for fascism,[71] intellectual critiques of planning[72] and attacks on the labour movement. Neville Chamberlain's 1938 journey to Munich is now routinely read as symbolic of all that went wrong during the 'low dishonest decade'.[73]

The spread of conflicts in which Europeans were involved during the 1920s and 1930s came to a catastrophic climax with the Second World War.

The fighting was geographically concentrated, mostly in Europe and East Asia. Other parts of the world were not directly affected by military activities, although they may have been combatants (USA, British dominions and colonies), or otherwise involved (Latin American countries were suppliers to the Allies). The Second World War resulted in great loss of life, the destruction of accumulated capital (public and private property), social dislocation, cultural shocks and political upheaval. At the start of the war the Europeans had worldwide empires, six years later the continent was ruined, occupied and divided.

The project of Britain in the post-war period

The situation of the British political-cultural project in the late 1940s was parlous.[74] The wealth of the country had been severely reduced, financial reserves were depleted and debts were owed to the USA. The physical infrastructure of the country was in degraded condition. The population was tired, yet in significant measure mobilized.[75] It was in these circumstances that the government of Clement Atlee began to renovate the project. Internationally, the context was taken to comprise three important spheres: the USA, the Commonwealth (created from the rapidly dissolving empire) and mainland Europe. The first strategy of managing this environment was Ernest Bevin's 'Churchill strategy': Britain was to sit within this triangle. At the same time domestically the project centred on the creation of the welfare state.

First moves I: machineries of accommodation (Bretton Woods, the UN and NATO)

A series of influences – centrally depression and war[76] – shaped US wartime policy-making and persuaded President Roosevelt and his allies that extensive economic and political reform within the global system was the key to securing the liberal-democratic system they favoured. A series of institutions were put in place, backed by the economic and political power of the USA: the Bretton Woods machinery,[77] the United Nations[78] and a little later (as the Truman regime gave ground to domestic pressures and allowed the cold war to develop) the apparatus of the North Atlantic Treaty Organization (NATO).

The Bretton Woods organizations included the International Bank for Reconstruction and Development (IBRD), the International Monetary Fund (IMF) and a little later the General Agreement for Tariffs and Trade (GATT) (1947). The Organisation for European Economic Cooperation (OEEC, becoming the OECD in 1961) was established in 1948 in Paris to coordinate the distribution of Marshall Plan assistance in Europe. These institutions were central to the US project to reconstruct a liberal capitalist trading sphere in the years after the Second World War, providing the framework within which European economies recovered. The United Nations, NATO and the rhetorical confection of cold war together provided the political counterpart. NATO was of particular importance, both as an institutional complement to the Bretton Woods apparatus but also as the vehicle of an elaborate ideological construct – 'the free West'. These institutions also reshaped the British political project; it was here in the years before Suez that the political elite positioned themselves. It was from these institutions that a flow of ideas ran into the domestic sphere (centring on the welfare state), shaping elite and subaltern perceptions, or political-cultural identities. The idea of the free West found distinctive subaltern expression within the heart of UK public politics as it melded with domestic ideas to produce a series of rhetorics of consolation amongst the British elite: the 'special relationship', the 'number one ally' and the community of 'English-speaking nations'.

These institutions had an initial orientation. They developed over time. Conventionally the economic apparatus is taken to have been eclipsed in the early 1970s when the financial pressures of the war in Vietnam forced the US to suspend the convertibility of dollar notes into gold, thereby precipitating an end to the successful post-war fixed currency rate system. A period of floating rates followed. The Bretton Woods machinery was reoriented to order the poor countries of the Third World, but their performance has been heavily criticised (in recent years debate has considered both their performance and the possibility of their radical reform[79]). The UN has had an analogous trajectory. As its membership and status has increased, so too has US displeasure. Finally, the apparatus of NATO lost direction after the end of the cold war in 1989/91; by the end of the interregnum in 2003 it was widely felt that the organization had no particular role to play.[80] On the back of these debates, a broader conversation began. As the Iraq crisis unfolded the very nature of the US/EU relationship was called into question, as was the British elite's continuing aspirations to a special status in Washington.[81]

First moves II: the machineries of continuity

The British polity has no formal constitution; rather it has a set of institutions that has accumulated over the period since the English revolution. Political and administrative power is concentrated at the centre in the institutional/ideological form of the sovereign crown-in-parliament.[82] It has been called the 'Westminster model'.[83]

The Westminster model

David Marquand[84] analyzed the polity as a project, a set of context-bound actions oriented towards particular goals. The structure of the polity could be looked at in terms of benefits and disbenefits or, looking to the future, capacities/incapacities. The British political system leaves power concentrated in the hands of the executive and power is protected against rival claims from local or regional government, from supra-national bodies, or from a written constitution. The state machine has an extensive reach in the form of official institutions and a spread of quasi-governmental organizations.[85] A particular element of the machinery is the nexus of Church, army and royal family. In general, recruitment to the establishment is secured by the public school/Oxbridge link.[86] These social-institutional locations embody and pass on patterns of identity and political-cultural identity. It is a very flexible system which is able to evade criticism, co-opt critics and shift and adapt to changing circumstances.

Marquand argued that changing UK circumstances required a variant of the 'developmental state',[87] which could serve as the vehicle for a novel British development project. In this way Marquand sketched the way in which a new route to the future might be envisaged. At the same time, he identified crucial inhibitions – in particular, the Westminster model. The sets of practices and ideologies that the Westminster model embodies and promotes are quite specific – elitist, secretive, patrician and conservative. The system could not undertake the task of ordering and mobilizing the population, the key to the success of developmental states, as it lacks both the capacity and the legitimacy.[88]

The Whitehall machine

The institutional structures of the British state are extensive. The polity has a dense armature of bureaucratic institutions which reach deep within society (machineries of discipline[89]). The core machineries are concentrated in and around London, and the heart of the apparatus is Whitehall/Westminster.[90] It is the permanent government. The idea of parliament plays a key role in the popular legitimation of the system, presenting a particular myth: its central directing role and power, its obedience to the electorate (via the party manifesto, elections and the receipt of a mandate) and its accessibility to ordinary people (via 'surgeries' that enable local politicians to 'represent their constituents').[91] The myth is dutifully rehearsed in the mass print/broadcast media.[92]

The role of the state is analyzed in a rather different idiom by 'regulation theorists',[93] whose work derives from French stucturalist Marxism.[94] It deals with the relationship of economic projects and political formations, using the ideas of modes of production and modes of regulation. There are a number of strands of work from this viewpoint: one has looked at advanced capitalist societies; another has looked at the Third World; and one strand has looked at cross-state class alliances.[95] The materials have also been used in the context of the British polity.[96] The machinery of the state services the needs of a particular economic formation. It also embodies and promotes distinctive sets of ideas and ideologies: an official national past[97] that is hierarchical, deferential and conservative. Yet there are structurally occasioned pressures for change. As the core countries of the mainland advance, the British polity looks evermore archaic.[98]

The palace

The monarchy plays a series of roles within the British polity: functional, as a part of the oligarchy, and symbolic, as the capstone of the political-cultural notion of 'Britain' and 'Britishness'. The monarchy is a central part of the establishment with links to the political class, senior bureaucracy, parliament, the armed forces, the Church, universities and professions. It also reaches down into the territories of the subaltern classes, with its links to civic, charitable and community organizations. The monarchy also plays a symbolic role, exemplifying timeless community and historical continuity. It is a key element in the official national past and national identity.

The monarchy embodies an elaborate set of ideas and ideologies: hierarchy, deference and continuity (conservatism). There are, however, pressures for reform: elite arguments for the reform of the institution (a bicycling monarchy on the northwest European model); arguments for its abolition (an understated republicanism is now evident in UK political discourse); arguments for its reclassification (from political to cultural – taking the Queen out of politics or the politics out of the Queen).[99] In addition there have been popular calls for change: a loss of sympathy for the family following the death of Princess Diana, routine calls for tax paying and criticisms of 'hangers on'. There are also patterns of reform that are implied by the project of the European Union – it has been pointed out, by way of illustration, that in the context of the federal political system of the United States it is not necessary to have a 'Queen of New Jersey'.[100]

The social networks of the establishment

The elite constitutes itself through elaborate social networks (schools, universities, clubs, business and so on), specific accents and voices (the ways in which social status is signalled), distinctive patterns of consumption (material signals of status), and preferred locations (the geography of desire, spatial segregation marking status).[101] The establishment embodies a particular set of ideas and ideologies about 'deepest Britain': tradition, discipline, service and achievement.[102] The project of the EU threatens their social and cultural position, as accumulated cultural capital is location specific. However, there seems to be no reason to suppose that the bulk of the elite could not individually 'shift into Europe', so one might anticipate an eventual creative accommodation: shifting class fractions (elite winners and losers) with new European social spaces in which to operate.

The post-war settlement I: the pursuit of economic growth

The planning machineries of the welfare state were the economic/ideological core of the post-war settlement.[103] A dual commitment was made to full employment and social welfare. These goals were pursued by all political parties in the post-war years until the confusions following the break down of the Bretton Woods system, plus oil price hikes in the 1970s, led to a problem with inflation and low growth which in turn paved the way for the neo-liberalism of the 1980s.

The pursuit of growth

The British government established a command economy in 1939–45. The population was mobilized for war production with import/export controls, financial controls, production/price controls, direction of labour and rationing. The experience of the command economy, plus the lessons of the work of J. M. Keynes, plus the demands of the population that the government avoid falling back to the situation of the 1930s, generated a disposition to pursue the objectives of growth and welfare.[104] This was done within the framework of the Bretton Woods system – economic prosperity was to be sought within the framework of a liberal trading sphere – however, the British system did have a distinctive corporatist aspect, the elite-level representation of functional groups.

The work of J. M. Keynes provided the rationale for a British version of corporatism. The elites of the state bureaucracy, business organizations and trades unions met routinely within the context of state-planning organizations in order to coordinate the governance of the economy and society. The economy was successful, with high employment, the establishment of the welfare state and, from the 1950s onwards, a rising consumer prosperity.

The confusions of the late 1970s, and the reactionary 1980s

The corporatist policy stance stayed in place until the slow decline of the Bretton Woods system and the impact of oil price rises generated a widespread doubt and confusion which was exploited by the theorists of neo-liberalism whose work provided the rationale for the shift to free-market policies. Over the 1980s and 1990s the corporatist orientation of the British state was revised. The legacy of the period is disputed. There was significant economic restructuring, extensive privatization of state assets, an end to elite-level corporatist meetings (that is, the representatives of labour were excluded) and financial liberalization. The late 1980s saw a credit-fuelled economic boom. Thereafter, there was slower growth, then further liberalization, deregulation and privatization. The years around the millennium saw a further credit fuelled economic boom, but all this was accompanied by increased inequality and insecurity (e.g. the dilution of employment protection, pension scandals and reforms).

The concern for security re-appears

The institutions of the Keynesian welfare state were strongly criticized by neo-liberals, and in the 1980s the system was rebalanced in line with their theories. The results were not widely credited with success. The economic policy decisions of Prime Minister Thatcher's government owed as much to opportunism and happy accident as they did to any clear ideology. None the less, the broad preference for 'market solutions' has marked the governments of her successors, John Major and Tony Blair.

The domestic debate has intermeshed with wider discussions. The future of the 'European social model' has been considered, so too the putatively over-riding logic of globalization.[106] The New Right's arguments were slowly submerged within a wider, superficially non-party, debate about 'globalization'; in the 1990s the notion provided a continuing rationale for the neo-liberal project.[107] It was however, vigorously contested.[108] As indicated, the notions of globalization are implausible; better to speak of internationalization and regionalization. The project of the European Union was similarly contested; to both proponents and opponents it represented an alternative to the British neo-liberal model.

Will Hutton[109] has argued that the neo-liberal lack of concern for economic security lies at the heart of contemporary economic problems. Hutton argues for a developmental capitalism, a variant of the Keynesianism that was put in place after the economic confusions of the 1930s and 40s. In the future the continuing concern for security may receive support from social-market/corporatist institutions on the mainland.

The post-war settlement II: project of the welfare state

The Beveridge Report provided the blueprint for the social welfare reforms of the 1945/51 Labour government. The reform programme constituted the social and ideological core of the post-war settlement – Beveridge spoke of five problems to be overcome on the road to reconstruction (want, disease, ignorance, squalor and idleness),[110] and offered a route forwards: (i) the provision of social housing; (ii) the establishment of a national health system funded through general taxation; (iii) the provision of similarly funded schooling for all children; and (iv) the establishment of a national social security system.[111] In the 1980s the New Right attacked the welfare state as inefficient, out-dated, over-loaded with demands and generating dependency within the population. However, its widespread popularity limited direct attacks, rather there were re-organizations to create internal markets and a new managerial ethos. In the early 1990s concern was expressed for the refunding and repair of the damages of the 1980s (and the long-term neglect of previous decades and governments). The reform of the welfare state became a key theme of the government of Prime Minister Blair.

Social housing/urban planning

In 1949 the government established a system of land-use planning. It was one of a series of measures designed to reorder the physical and social capital of the country. Local and regional governments were required to publish structure plans detailing expected urban growth. The central government initiated major schemes for new towns, roads, airports and so on. A key area of action was in housing. A series of successful phases can be identified that began with the post-war slum clearance and the repair of war-occasioned deficits in housing provision. This was followed by the increased drive to provide housing in the sixties that took the form of system-built tower blocks that turned out to be vertical slums.[112] The status of social housing altered – from good provision for all to low-quality provision for those who had absolutely no alternative. The great sell-off of the 1980s confirmed this view. It persisted until the mid-1980s collapse of the housing market reminded everyone that social control and provision of housing was necessary in a modern society. In the 1990s a renewed economic expansion lead to further changes in the housing market – in particular new sources of private rented accommodation – and in the early years of the new century a further housing price boom took place on the back of a buoyant economy and very low interest rates. In the context of UK economic and social life, housing plays a very significant role – in terms of provision (social/private), location (greenfield/brownfield, and the growth of the southeast), economy (where housing, now mostly private, is a source of savings/capital gains – and speculation) and symbolic meaning (where private was positive, social was subaltern). It remains a core concern of government.

Social health

The National Health Service (NHS) was established to provide comprehensive medical care free at the point of treatment and funded through general taxation and a contributory national insurance scheme. It was a success, and has received popular support over the entire period of its existence. It escaped direct attack from the New Right in the 1980s but came under routine low-level attack. The government of Prime Minister Blair made more money available as it became clear that the system had been chronically under-funded (in comparison with mainland systems) for years (familiar claims to inexpensiveness and efficiency gave way to the admission that more money was indeed needed).

Schooling

The school system has been the institutional location for a series of conflicts over the entire post-war period – in effect, elite versus mass, with the aspirant middle classes as the 'swing voters'. It is a site of contested meanings: for middle-class parents private schooling is often read as better than state schooling and a host of judgements about social status revolve around these characterizations.

Social welfare

The social welfare system has been a success when compared with pre-war provisions, and was generally supported in the period of the long post-war economic boom. However, in the 1980s the New Right attacked the system. It is a site of contested meanings – community provision versus private provision – and once again a host of judgements about social worth and responsibility revolve around these ideas.

Overall, we can note that the machineries of the welfare state remain popular, and that the values they embody, often summarized around an ethic of 'fairness', continue to receive support. A reordering of the welfare system in line with the northwest European model would ensure the continuation of the system and its ethic.[113]

The post-war settlement III: dynamics of popular cultural change

The idea of culture is contested, and unpacks in a variety of ways.[114] Zygmunt Bauman[115] speaks of culture as praxis. A clutch of concepts deal with the routines of life: popular culture, mass culture, urban culture, consumer culture and the culture of everyday life. They point (rather differently) to the sets of ideas carried in everyday social practice; the means to self-understanding and self-expression, possibly unsophisticated, but none the less widely available.

The realm of popular culture can range from folk arts through formal arts and literature and into the commercial realm. It is a dense sphere of activities and meanings. The sphere of popular culture can generate self-conscious reflection upon the patterns of life, whilst at other times it merely provides a vehicle for commercial products (passive and/or expressive consumption[116]). The materials of popular culture are drenched in history, a point made by Hoggart and Hall.[117] The contemporary period is rooted in the experience of the war years and changes down the years; there is a wealth of available popular cultural material. It can be summarized chronologically: the affirmations of the 1940s; the critical work of the 1950s; the changes of the 1960/70s; the reactionary debates of the 1980s; and the unsettled consumerism of the millennium. The debates and concerns of each period carry forward into the debates and concerns of the next period. This chronological treatment will bring us to the issue of likely patterns of change in economic, social and cultural patterns with the movement towards a unified Europe.

The 1940s – from wartime to never again

The experience, memories and myths of the Second World War inform one popular version of Britishness. The key is the idea of the 'wartime'. It is a shared experience of community spirit, collective endeavour, shared danger and shared sacrifice. The long period of danger and upheaval led to a final morally validating victory. It is a period when action made a difference,[118] available to be invoked thereafter in a series of social situations, ranging from football matches through financial crises to further wars.

The period of wartime produced the commitment to the goals of growth and welfare, noted earlier, along with a determination on the part of the population to hold the political elite to their promises and these goals.[119] The informal slogan was 'never again' and Peter Hennessy[120] takes this as the tag for the sets of ideas of the reform period that saw the post-war settlement put in place.

By the end of the 1940s many of the patterns of debate that were to recur one way or another over the remaining years of the short twentieth century were in place: the commitment to a welfare state, underpinned by Keynesian corporatism;[121] and cutting against this the resistances of the old ruling groups which, by the last years of the decade, found official expression in what was to develop into the whole cultural apparatus of the cold war with its related subsidiary conceit of the alliance of the English-speaking peoples.

The early consumerism and proto-rebellion of the 1950s

The post-war period saw employment, security and health for the majority of the population. The 1950s offered a first experience of consumerism for many people, and a new sphere of domestic consumption and leisure developed. The exemplar of consumerist life-style was the USA. The novelties were not simply products; they were practices (new highways, out-of-town supermarkets, brand names, urban centre redevelopments, retail/leisure intermixing and so on) and the idea of a new pattern of life – affluence.[122]

The 1950s saw the political elite involved in the rich development of the apparatus of the cold war (alliances, military organization, defence, propaganda etc.). After the 1956 Suez debacle they confronted the end of residual dreams of empire and embraced the notion of a 'special relationship' with the USA. Looking at this relationship, Hitchens[123] has diagnosed an elite-sponsored fawning self-delusion which presents the Anglo-American alliance as the core of the globally privileged realm of the English-speaking peoples. In this way the British ruling class could pretend to influence in a vicarious continuing empire, whilst the Americans could be subtly seduced by the self-same archaic and anti-democratic vision of superiority and centrality. The wider sphere of elite culture seemed unchanged: opera, the home service radio, sports meetings (Wimbledon, Lords,[124] Cowes Week, Henley etc.), country houses[125] and the social networks of town and county – the continuing social world of the establishment. However, the growing material wealth and the decline of deference produced cultural rebels (and moral panics): teddy boys, rock and roll, the beat generation and the angry young men with their attacks on the alleged[126] – 'you've never had it so good' complacency of the years of Harold Macmillan's premiership.

The cultural revolution of the 1960s

The post war long boom ran through the 1950s and 60s. There was sustained full employment; a new consumer culture emerged; the political elite, led by the government of Prime Minister Alex Douglas Home, came to be seen as out of touch. There were major public scandals: the Profumo affair; and the trial of the publishers of the D. H. Lawrence novel *Lady Chatterley's Lover* (with the prosecuting council famously asking one witness if he would 'allow his wife or servant' to read the book). In the popular arts there was a satire boom, and politicians were favourite targets. The accession to power of Prime Minister Harold Wilson was widely taken to mark a significant new departure for the people as a whole.

The 1960s saw the widespread dissemination of culture-critical material. A self-consciously modernizing government, economic prosperity, social liberalization and an unexpected flowering of the arts all contributed to a period of cultural renewal. Hanif Kureishi[127] has remarked that the Beatles encouraged artists, and social groups generally, to develop their own talents rather than waiting for permission and approval from established arbiters of cultural and social/political taste. At the same time however are suggestions that the government of the day failed to achieve its goals – it is a mixed legacy. The episode has come to be somewhat implausibly demonized by the political right.

The post-war settlement revisited

The post-war commitment to the state-sponsored pursuit of domestic growth and welfare within the wider international context of the US-led liberal trading area came under pressure in the late 1970s: war, petro-dollars and stagflation. An available counter-argument to the post-war settlement, the arguments of the neo-liberals, first voiced in the war years,[128] was translated into global practice. In the UK the experiment became personalized: 'Thatcherism',[129] an authoritarian populist market-oriented reform.[130] Prime Minister Thatcher's project unfolded via a series of specific policies – some planned, some opportunist; some successful, others not – overall an uneasy mix of neo-liberal and neo-conservative ideas.

'Thatcherism' has been characterized as a mix of neo-liberal economics and social authoritarianism: the former is an ideological construction that expresses a quite particular view of humankind in respect of the matter of securing a livelihood; the latter is a readily available British cultural resource. The period included economic growth, 'yuppies', 'sleaze' and privatizations. There was the repression of organized labour (via a slate of new laws and one crucial victory over striking coal miners). There was also a denial of reason in favour of convenient ideology, evidenced, for example, in the use of certain key phrases – the 'economic arithmetic', 'there is no alternative', 'there is no such thing as society', 'got on his bike' and so on. The government of Prime Minister Thatcher offered a mixture of neo-liberal economics coupled with an authoritarian social conservatism which harked back to earlier ideas of Britishness – the package has been described as Thatcher's version of Churchill's idea of Britain, in brief, a nationalist fantasy. Stuart Hall correctly saw that Thatcher was not offering a simple reordering of familiar policies but instead a completely new political project. It is true that its policy expression was initially inchoate and it is true that it developed fortuitously, but her government set out to change the trajectory of British politics and in this they succeeded. The crisis for Thatcher turned out to be reconciling domestic aspirations with the unfolding logic of involvement in Europe – one might say that she recognized the threat, mis-characterized it and recoiled – others within the political class evidently took the view that effective strategies of delay and accommodation were available. The record of her government is debated.[131] Prime Minister Major followed her trajectory, and the late 1990s saw a measure of continuity – one commentator spoke of 'Blaijorism'.[132] The early hopes invested in the New Labour government following its 1997 electoral success were unfulfilled as nothing much happened: no disasters, no successes, indeed not much of anything. A new word entered the political language – 'spin' – referring to the government's desperation to be seen at all times in the best of all possible lights. A second landslide electoral victory followed; this time on the lowest recorded turn out in British political history. The placid pattern continued until it was radically disturbed by the 2003 Iraq war.

The early years of the new millennium were dominated by the implications of the collapse of cold war certainties: a leisurely debate following 1989/91, more urgent after 2003. The central issue has been Europe. Neal Ascherson[133] and Christopher Harvie[134] argued to the specificity of the experience of the Scots and unpacked the multi-national aspects of the United Kingdom. Europe is a new context for the Scots (or as the Scottish National Party had it, 'independence in Europe'). Gwyn Williams[135] traced the routine reinvention of Wales down the years and argued that Europe might just rescue such peripheral nations from eclipse-via-absorption, offering a better way of managing their relationships with their immediate state unit (Wales/Britain). In England there has been a very limited debate about the implications of the developing European Union. The political classes are broadly unsympathetic and the requirement to fashion new identities for new circumstances has not been publicly acknowledged.[136]

New directions: available resources and the demands of change

In the UK polity there are contending class groupings. In the years after 1945 there was a pattern of balance plus associated common understanding, tagged 'the post-war settlement' or 'consensus'. It was always a contested compromise and it was never fixed; it was always shifting along lines of conflict. It is not fixed at the present time: elite and mass interact; there are claims and counter-claims. However, structural argument can foster an overstated pessimism: elite aspirations to secure authority and discursive closure are routinely resisted and reworked; they are not mechanically transmitted and efficiently secured. Parkin and Hall have illuminated the forms these engagements might take. In the years following the Second World War the subordinate classes did secure significant economic and social reforms, yet much of the apparatus of the political system was left untouched. In time, as events unfolded, the machineries of the unreformed state were available to incoming neo-liberal groups and the machineries of the state were deployed to recentre the post-war settlement. Andrew Gamble[137] spoke of the strong state being used to make a free economy (the neo-liberal package centred on deregulation and liberalization, with privatization as a one-off corollary). However, two things are now clear: first, the structural context within which the British state-regime operates has radically changed – in the wake of the events of 1989/91–2003 it is a new game; and second, there are resources available to reconstitute a notion of Englishness.

Institutions, ideas and the dynamics of change

Institutions embody ideas about the nature of the community within which we live. These institutions (in part) tell us who we are and where we might expect to be in the future. Institutions do not determine identity, but they provide a framework within which individuals must move.[138] As might be expected, they are sites of sharp political conflict. In a period of relatively stable contested compromise the conflict may be muted or ritualized (the post-war consensus). However, if the sets of structural circumstances that enfold any collectivity begin to change, then as groups determine their reactions we might expect to find conflict in respect of the functions, embodied ideas and projects of familiar institutions.

The post-Second World War period has seen the machineries of political continuity stage a long and successful rearguard action in the face of the efforts of the subordinate classes to secure welfare reform in the wake of the confusions of the Second World War. This success finds evidence in the episode of the post-war consensus. As events continued to unfold, by the end of the reactionary period of the 1980s, which saw a sustained attack on the institutions and ethics of the welfare state, the elite might have expected a period of relative material advance, enhanced social visibility and political-cultural influence. However in the period 1989/91 the wheel of history turned and the new centre of gravity of the politics of the UK and its institutional structures became the developing organization of the EU. As the general political tenor of mainland Europe reflects a different historical development experience, which might be summed-up as social/Christian democratic,[139] the prospects for the continued domestic pre-eminence of the British elite are diminished.

The British state-regime now inhabits a new context – in the wake of the events of 1989–2003 the following points must be granted: (i) the global system is now significantly internationalized – trading networks span the globe – but there is no single integrated globalized system, this is a neo-liberal myth, a recycling of 1950s/60s 'convergence theory'; (ii) the global system evidences distinct signs of regionalization – it is possible to make plausible arguments to the effect that the global system is in some measure now tri-regional; and (iii) the immediate non-negotiable, unavoidable given context for the British state-regime is the project of the European Union – a threat to the elite and an opportunity to the masses.

The vitality of available resources considered

The political-cultural project of Britain/Britishness has its historical occasion in the late eighteenth century. It reached a peak in the years before the Great War. It experienced an Indian summer in the years between the wars. It has adjusted to a series of severe shocks over subsequent years. In recent times it has been a subordinate part of the American liberal-trading sphere. The domestic political-cultural expression of these circumstances is a mix of elite consolations and subaltern acquiescence.

It is also an historical achievement. It is stable. It is also contingent. It is a contested compromise, liable to change as circumstances change. It is clear that over the period 1989/91–2003 circumstances changed – domestic change is thus inevitable. The structures of the global system do not stop at the official state border, rather they run through the domestic economy and society. Structural change has entirely non-negotiable impacts (these structures enfold the patterns of our routine lives) and elites and masses cannot do other than react: novel understandings and new choices.[140] A new contested compromise might be anticipated.

The political-cultural construct of Britain/Britishness is a top-down official national identity, not a popular democratic national identity. It was engineered and its needs must be routinely re-engineered (there are 'machineries of universe maintenance'[141]). In other words, the population may acquiesce in the received political discourse but they are not passive – the official discourse has to be made and remade. One might ask, therefore, just what non-elite ideas are available, and how might they be put to use during a period of relatively abrupt complex change. What ideas are there, in other words, with which an alternative national identity might be imagined – an alternative route to the future?

We start from where we are: (i) inherited structures (ideas and practices, including the resources of the 86 per cent of the population of the Isles who live in England – demobilized, but not depoliticized; in other words, politics continues in NGOs, charities, pressure groups, the media, local organizations, popular opinion, gossip and so on); (ii) the new context of internationalization and regionalization; and (iii) the particular demands and opportunities provided by the project of the EU. As regards the first key area, in addition to the continuing demands of the official ideology we can note that the social-institutional arrangements of Keynesian corporatism and Beveridgean welfarism command extensive support amongst the population. The circumstances of internationalization and regionalization offer new resources – trade, travel and knowledge. The sphere of the EU offers a stock of 'good examples': we can compare domestic arrangements with those of our neighbours – an obvious strategy and one likely to carry conviction, more immediately intelligible than the routine recourse to American models made by the elite. All these can be taken to comprise a reserve of available resources that could be drawn upon – in conjunction with newly opening structural changes within Europe, and thus the territory of the Isles – to fashion a reconstituted England/Englishness.

The new space of the European Union

The UK population inhabits an oligarchic system[142] which reserves power unto itself (i.e. the establishment), and routinely and assiduously acts to demobilize the population (who are invited to be modest, civil, polite and await the due pace of elite reflection upon their complaints, claims and aspirations). None the less, there is vigorous subaltern activity (else the continuing effort of demobilization would not be necessary). The settlement/consensus is a contested compromise, it has to be made and remade in routine social life. The space of the EU is one further arena of contestation – at present largely ritualized amongst political elite fractions (Eurosceptics and pro-Europeans) with the mass of the population seemingly disengaged or resigned.

As noted, the structural circumstances enfolding the country are captured, or controlled, via a series of institutional locations. The machinery of the state enables the ruling regime to read and react, to contrive projects and routes to futures. The domestic counterpart entails order and legitimation – it is in all a complex package, a contingent achievement. In the UK polity there are various institutional locations which disseminate ideas and ideologies in respect of power and legitimacy: parliament, monarchy, the machineries of the welfare state and so on. These are contested sites: first, the Keynesian welfare state, and more recently the project of neo-liberalism; it may well be that the political elite judged that it could continue with its Atlanticist neo-liberal package – thus, 'Blaijorism'[143] – but the events of 1989/91–2003 are outside its control. It is my speculation that these sets of structural changes will impact upon domestic institutional machineries and ideas. Change will oblige the ruling class to respond and these responses will have to be made within a situation which is novel (it always is). Today the global economy is internationalized, there is significant regionalization and dominating the UK situation there is the available political project of the European Union.

In this context the contribution of the project of the EU – itself contrived in response to complex patterns of global and regional structural change – is important. The project of the EU offers a new political-cultural territory. It is a political space in the process of being shaped. It offers a coherent northwest European model of advanced industrial capitalism that is distinct from the liberal free-market system embraced by the British. It is within the political-cultural space of the developing European Union that a notion of England/Englishness might be relocated.

As these complex dynamics unfold there will be new occasions for elite and mass to contest 'the national project'. It seems probable that one area of contestation will be within the institutional machineries of the polity; another will be in the realm of ideas and identity, the projects the polity affirms. At this point we can turn away from history, institutions and ideas and look more specifically at political discourse. While the foregoing material offered sequential or diachronic analysis, we now turn to synchronic analysis, considering the ways in which the polity works. Recalling Peter Winch[144] and Antonio Gramsci,[147] just what are the sets of ideas running through people's heads? How are ideas of 'Britishness' made and remade and where are the resources for a re-expressed 'Englishness'?

Notes

1. This chapter is derived from P. W. Preston 1994, *Europe, Democracy and the Dissolution of Britain*, Aldershot, Dartmouth, chapters 2, 3 and 4.
2. L. Colley 1992, *Britons: Forging the Nation 1701–1837*, Yale University Press.
3. C. Bridge and K. Fedorwich (eds) 2003, *The British World: Diaspora, Culture and Identity*, London, Frank Cass.
4. Elsewhere I called these a 'general theory'; it is a summary statement rather than the outline of an empirical research programme, and points to theorizing as engaged and critical. See P. W. Preston 1996, *Development Theory*, Oxford, Blackwell.
5. E. Gellner 1983, *Nations and Nationalism*, Oxford, Blackwell.
6. E. Gellner 1964, *Thought and Change*, London, Weidenfeld.
7. B. Anderson 1983, *Imagined Communities*, London, Verso.
8. T. Nairn 1988, *The Enchanted Glass*, London, Radius. I am using Nairn's work to sketch the historical occasions of extant institutional and ideological structures. On the work of Nairn see the foreword to P. Anderson 1992, *English Questions*, London, Verso.

9 N. Davies 2000, *The Isles: A History*, London, Papermac, pp. 353–426.

10 See R. Porter 2000, *Enlightenment: Britain and the Creation of the Modern World*, London, Allen Lane.

11 See Colley 1992.

12 N. Davies 1997, *Europe: A History*, London, Pimlico, pp. 432–3, 382.

13 Davies 2000. This is a key point, somewhat lost in the detail of the text.

14 The title of a briefly influential anthropology of a Mexican shaman, including much on drug-induced dreams and the inability of the modern mind to comprehend them. See Carlos Castenada 1971, *A Separate Reality*, London, The Bodley Head. In a similar vein on England, see R. Scruton 2001, *England: An Elegy*, London, Pimlico.

15 I began thinking about the simple geography of the archipelago when I realized that when Ulstermen spoke of the 'mainland' they meant the main island, not continental Europe – a wonderful example of ideology over-riding both simple geography and English usage.

16 Davies 2000, p. 426.

17 Barrington Moore 1966, *The Social Origins of Dictatorship and Democracy*, London, Allen Lane.

18 Gellner 1983.

19 Anderson 1983.

20 I. Buruma 1999, *Voltaire's Coconuts, or Anglomania in Europe*, London, Weidenfeld.

21 H. Kaye 1984, *The British Marxist Historians*, Cambridge, Polity.

22 Porter 2000.

23 Buruma 1999.

24 D. Canadine 1998, *Class in Britain*, Harmondsworth, Penguin.

25 Nairn 1988, pp. 151–2.

26 Buruma 1999.

27 Detailed by S. Schama 2003, *A History of Britain III: The Fate of Empire 1776–2000*, London, BBC.

28 Nairn 1988, p. 154.

29 Roy Porter (2000) celebrates the robust commercial vigour of the period.

30 Bridge and Fedorowich (2003) provide a way of reading the diaspora. See also D. Canadine 2001, *Ornamentalism: How the British Saw their Empire*, London, Allen Lane. He argues that the idea of Britain seems to emerge on the periphery of the polity and that the English elite seem content to eschew a modern English nationalism to avoid links to democracy.

31 Davies 2000.

32 Buruma 1999.

33 Colley 1992.

34 Colley 1992.

35 On this see Schama 2003. He shows that the reaction in the UK elite was ferocious, and that the image of the blood-stained French revolution in contrast to peaceable Britain is false – the truth was in fact the other way about.

36 An argument made by E. P. Thompson 1968, *The Making of the English Working Class*, Harmondsworth, Penguin.

37 R. Colls (2002, *Identities of England*, Oxford University Press) suggests that subaltern social, workplace and educational activity followed the political defeat of Chartism.

38 Nairn 1988, p. 245.

39 Anderson 1983.

40 Anderson (1983) makes the point that Britishness, like Russianness or Indianness, was imposed by the elite on their subordinate population.

41 See E. Hobsbawm and T. Ranger (eds) 1983, *The Invention of Tradition*, Cambridge, Canto.

42 Thompson 1968.

43 Nairn 1988 p. 215.

44 P. Calvacoressi 1997, *Fall Out: World War II and the Shaping of Postwar Europe*, London, Longman, p. 135.

45 D. Canadine 1983, 'The Context, Performance and Meaning of Ritual: The British Monarchy and the Invention of Tradition, 1820–1977', in Hobsbawm and Ranger (eds). In regard to the empire see Canadine 2001. Interestingly Scruton (2001) stresses the idea of 'home' in the construction of England/Englishness, adding that it was how those overseas thought – class mores exported and then re-imported?

46 Nairn 1988, p. 280.

47 Nairn 1988, pp. 280–1.

48 D. Marquand 1988, *The Unprincipled Society*, London, Fontana.

49 H. Overbeek 1990, *Global Capitalism and National Decline*, London, Allen and Unwin.

50 J. Saville 1988, *The Labour Movement in Britain*, London, Faber. Saville comments 'it is one of the remarkable characteristics of Britain during the twentieth century that ... the propertied groups, and their political representatives, have retained their economic and political power unimpaired' (p. 112).

51 Nairn 1988, p. 373.

52 A. Mayer 1981, *The Persistence of the Old Regime*, New York, Croom Helm.

53 Nairn 1988, p. 374.

54 A. Mayer 1988, *Why Did the Heavens Not Darken: The 'Final Solution' in History*, New York, Pantheon.

55 E. Hobsbawm 1994, *The Age of Extremes: The Short Twentieth Century 1914–1991*, London, Michael Joseph.

56 For a particular evocation, see J. G. Ballard 1988, *Empire of the Sun*, London, Grafton.

57 Davies (1997) points out that the slide to war began in 1938 with Austrian *Anschluss*; thereafter the 1938 Munich crisis doomed Czechoslovakia which was occupied in early 1939.

58 Mayer 1988.

59 Mayer 1988; Mark Mazower 1998, *Dark Continent; Europe's Twentieth Century*, New York, Alfred Knopf; Christopher Thorne 1986, *The Far Eastern War: States and Societies 1941–45*, London, Counterpoint. Data on the European death tolls is given in Davies 1997.

60 See S. Strange 1988, *State and Markets*, London, Pinter. In her terms, for Western Europe, productive, financial, security and knowledge power shift across the Atlantic.
61 There were many debates about the collapse of civilization – how could it happen, who was to blame. In regard to the latter, an answer was given in the Versailles Treaty in a clause on 'German war guilt'.
62 See E. H. Carr (1939) 2001, *The Twenty Years Crisis*, London, Palgrave.
63 J. K. Galbraith 1975, *The Great Crash*, Harmondsworth, Penguin.
64 There are further versions of fascism available in the USA, Japan and the China of the KMT (the Chinese Nationalist Party).
65 Associated with Neville Chamberlin, but advocated by E. H. Carr. In a related way J. M. Keynes argued that the German people should not be punished for the Great War and in the 1930s there was a strong anti-war movement. It was not until the late 1930s that British defence spending was increased.
66 A series of social scientific explanations of fascism has been offered: psychological – mad people caused it all; international relations – the unjust end to Great War generated fascist reaction; political history – the circumstances of interwar Europe were poor and fascism emerged as the creed of the resentful provincial lower-middle classes, and once the project was up and running it developed its own mad logic and ran out of control; Marxist political economy – fascism is an extreme variant of capitalism; Marxist cultural history – fascism is an extreme expression of modernity; and historical-materialist – fascism was an element of a general crisis.
67 Schama 2003, p. 335 – houses were built but the early efforts soon faded.
68 Mazower 1998.
69 Mazower 1998.
70 See P. Addison 1977, *The Road to 1945*, London, Jonathan Cape.
71 R. Griffiths 1983, *Fellow Travellers of the Right*, Oxford University Press. See also K. Ishiguro 1989, *The Remains of the Day*, London, Faber and Faber.
72 For example, F. Hayek 1944, *The Road to Serfdom*, London, Routledge'; K. Popper 1957, *The Poverty of Historicism*, London, Routledge.
73 W. H. Auden's poem, *September 1st 1939*.
74 One German colleague remarked, 'They lost more than we did' (Prof. Dr Gunter Schlee, personal communication). The damage to both countries was severe, and the British idea of 'winning the war' is mostly myth. In fact, the island was a base for the USA, the Red Army did most of the fighting and the war saw the global role of Britain destroyed. The scale of the 1939–45 British catastrophe is not often remarked on, but for shocking numbers, see Davies 1997.
75 Addison 1977.
76 G. Kolko 1968, *The Politics of War*, New York, Vintage.
77 On the debates, see R. Skidelsky 2002, *John Maynard Keynes, A Biography: Fighting for Freedom* (Vol 3), London, Penguin.
78 T. Hoopes 1997, *FDR and the Creation of the UN*, Yale University Press.
79 J. Stiglitz 2002, *Globalization and its Discontents*, London, Allen Lane.

80 S. Croft, J. Redmond, G. Wyn Rees and M. Webber 1999, *The Enlargement of Europe*, Manchester University Press

81 R. Braithwait 2003, *Prospect* March/April.

82 Colls 2002, chapters 3 and 4 details the construct.

83 Marquand 1988.

84 Marquand 1988.

85 R. Miliband 1973, *The State in Capitalist Society*, London, Quartet. See also R. Miliband 1972 (2nd edn), *Parliamentary Socialism*, London, Merlin; R. Miliband 1984, *Capitalist Democracy in Britain*, Oxford University Press.

86 J. Paxman 1990, *Friends in High Places*, Harmondsworth, Penguin.

87 An idea taken from discussions of the success of East Asia; it has deeper roots in late-nineteenth-century German economic theorizing. See Chalmers Johnson 1992, *MITI and the Japanese Miracle: The Growth of Industrial Policy, 1925–1975*, Tokyo, Tutle; Linda Weiss 1999, *The Myth of the Powerless State: Governing the Economy in a Global Era*, Cambridge, Polity.

88 There is a vast literature on the experience of the East Asian developmental states. For current debates (earlier versions revolved around Germany in particular, e.g. Friedrich List as the theorist of 'late development') see Chalmers Johnson, Robert Wade or Linda Weiss, all of whom argue for the particularity of the development experience of East Asia, and all of whom identify a key role for the state and their projects of national development. Marquand's point is that the state must mobilize its population with a mix of capacity and legitimacy – the Westminster model has neither.

89 See for example Michel Foucault 1977, *Discipline and Punish: The Birth of the Prison*, New York, Pantheon.

90 P. Hennessy 1989, *Whitehall*, New York, Free Press. See also Anthony Sampson 2004, *Who Runs this Place? The Anatomy of Britain in the 21st Century*, London, John Murray – in particular the rather nice diagram on the first couple of pages of the book.

91 Miliband 1972.

92 It turns up in the radio soap *The Archers* in the form of the Parish Council, presented as a sort of exemplificatory grass-roots democracy.

93 Miliband had a sharp debate with Nicos Poulantzas; the structuralist work was disposed to abstract theorizing whereas the theorists located in the UK around this time were more disposed both to empirical research and the political work of the UK labour movement, dominated of course by the Labour Party. On all this see M. Newman 2002, *Ralph Miliband and the Politics of the New Left*, London, Merlin, pp. 198–215. On Miliband's treatment of the Labour Party's obsession with parliament, see pp. 73–7.

94 The 'French regulationist school' was informed by the work of Louis Althusser. One strand of their work looked at modern society, another at the Third World. See A. Brewer 1980, *Marxist Theories of Imperialism*, London, Routledge, on P. P. Rey.

95 The 'Amsterdam regulationist school' – Overbeek 1990 and K. van der Pijl 1984, *The Making of an Atlantic Ruling Class*, London, Verso.

96 See B. Jessop, K. Bonnet, S. Bromley and T. Ling 1988, *Thatcherism: A Tale of Two Nations*, Cambridge, Polity; C. Hay 1996, *Re-stating Social and Political Change*, Open University Press.

97 P. Wright 1985, *On Living in an Old Country*, London, Verso.

98 A point made by Nairn. It is also interesting to watch him edging towards discussing the actual project of the EU.

99 The government of Prime Minister Blair has moved to remove hereditary legislators from the upper house of parliament, questioning thereby the hereditary principle.

100 Earl Smith (personal communication).

101 See Sampson 2004, Paxman 1990.

102 On conservatism see Scruton 2001; also Ted Honderich 1990, *Conservatism*, London, Hamish Hamilton, who reduces it to a celebration/defence of selfishness.

103 The post-war settlement can be read as a *provisional contested compromise between contending class groupings*. There was a settlement, but there was no broad settled agreement or consensus. One can identify two clear breakdowns in what consensus there was: the Wilsonian turn to economic planning (K. Middlemas 1979, *The Politics of Industrial Society*, London, Andre Deutsche) and Thatcher's turn to neo-liberalism (S. Hall 1990, *The Hard Road to Renewal: Thatcherism and the Crisis of the Left*, London, Verso). The settlement was in great measure sustained, but it came to an end in 1989/91 when the wider global conditions that shaped it changed decisively.

104 Addison 1977.

105 P. Kerr 2001, *Postwar British Politics*, London, Routledge.

106 For UK celebrants see A. Giddens 2002, *Runaway World: How Globalization is Shaping our Lives*, London, Profile; D. Held and A. McGrew 2002, *Globalization/Anti-Globalization*, Cambridge, Polity.

107 See Giddens 2002 and Held and McGrew 2002.

108 See P. Hirst and G. Thompson 1999 (2nd edn), *Globalization in Question*, Cambridge, Polity.

109 W. Hutton 1996, *The State We're In*, London, Vintage; W. Hutton 2002, *The World We're In*, London, Little, Brown.

110 P. Addison 1995 (2nd edn), *Now the War is Over: A Social History of Britain 1945–51*, London, Pimlico, p. 10.

111 P. Hennessy 1992, *Never Again: Britain 1945–1951*, London, Jonathan Cape.

112 In Britain the dreams of modern architects about patterns of life and the use of new concrete forms did not meet with evident unrestricted success – rather the reverse. Interestingly, similar housing has been provided in cities in East Asia (Hong Kong and Singapore, for example) where it is successful; in the case of Singapore one reason could be found in social discipline.

113 The future of the European social democratic model has been widely discussed – see C. Pierson 2001, *Hard Choices: Social Democracy in the 21st Century*, Cambridge, Polity.

114 C. Jenks 1983, *Culture*, London, Routledge.

115 Z. Bauman 1973, *Culture as Praxis*, London, Routledge.
116 F. Jameson 1991, *Postmodernism: Or the Cultural Logic of Late Capitalism*, London, Verso. Postmodernists have analysed the ephemera of commercial/media culture in terms of surface effect, depthlessness, pastiche and irony.
117 R. Hoggart 1958, *The Uses of Literacy*, Harmondsworth, Penguin; Hall 1990.
118 Wright 1985.
119 Addison 1977.
120 Hennessy 1992.
121 Addison 1977.
122 J. K. Galbraith 1958, *The Affluent Society*, Harmondsworth, Penguin.
123 C. Hitchens 1990, *Blood, Class and Nostalgia: Anglo-American Ironies*, London, Vintage.
124 On cricket see Mike Marquese 1994, *Anyone but England: Cricket and the National Malaise*, London, Verso.
125 Their decline and rescue took place in the 1950s when the National Trust was re-focused on 'saving' the large country houses which were falling out of use – see Wright 1985. For a rebuttal of critique of 'heritage' see Raphael Samuel 1994, *Theatres of Memory Vol 1: Past and Present in Contemporary Culture*, London, Verso.
126 After the debacle of 1956 Macmillan appreciated that empire was at an end, hence the famous 'winds of change' speech that prefigured the withdrawal from colonial Africa and elsewhere.
127 It is true that the 1960s did encourage experimentation – Kureishi has pursued this in film and novels.
128 See Hayek, Popper, Mont Perlin Society etc. (Hay 1996, pp. 130–6).
129 Hall 1990.
130 A. Gamble 1988, *The Free Economy and the Strong State*, London, Macmillan; also Jessop *et al.* 1988.
131 Kerr 2001.
132 The expression 'Blaijorism' comes from C. Hay 1997 'Blaijorism: Towards a One Vision Polity?', *Political Quarterly* 68, 372–9. See also C. Hay 1999, *The Political Economy of New Labour*, Manchester University Press. Hay discusses in detail the first 'New Labour' government with their relentless trope of 'new-ness', and comes to the conclusion that they looked to be little more than a vote-gathering apparatus, standing for nothing very much.
133 N. Ascherson 1988, *Games with Shadows*, London, Radius.
134 C. Harvie 1992, *Cultural Weapons: Scotland and Survival in the New Europe*, Edinburgh, Polygon.
135 G. Williams 1985, *When Was Wales?* Harmondsworth, Penguin.
136 Discussed in P. W. Preston 1994, *Europe, Democracy and the Dissolution of Britain*, Aldershot, Dartmouth.
137 Gamble 1988.
138 A. P. Cohen 1994, *Self-Consciousness*, London, Routledge.
139 Marquand 1988.
140 After Peter Winch, as we understand, we change; after Alasdiar MacIntyre, as we understand, we choose. P. Winch 1958 (2nd edn 1990), *The Idea of a Social Science*

and its Relation to Philosophy, London, Routledge; A. MacIntyre 1962, 'A Mistake about Causality in Social Science', in P. Laslett and W. G. Runciman (eds), *Philosophy, Politics and Society, Series 2*, Oxford, Blackwell.

141 The expression comes from P. Berger and T. Luckmann 1967, *The Social Construction of Reality*, London, Penguin. See part 2, chapter 2.

142 A point made by Calvocoressi (1997, p. 135): '[The UK] may plausibly be labelled an oligarchy tending towards democracy ... its rhetoric and some of its institutions are democratic, but its values and habits remain significantly oligarchic.'

143 Hay 1997.

144 Winch 1958.

145 Who gave us the idea of hegemony – A. Gramsci 1971, *Selections from the Prison Notebooks*, edited and translated by Q. Hoare and G. N. Smith, London, Lawrence and Wishart.

5 British political discourse[1]

The global system might be thought of as a contingent, shifting pattern of relationships, available to any one particular agent (person, group or state) in terms of the notion of structure (the sum total of what every other agent does, the way the extant pattern of relationships presents itself to the particular agent, the given). The state can be thought of as a membrane controlling trans-state relationships/structures. The business of the state is thus to accommodate these trans-state relationships, to grasp and utilize them in pursuit of its own project. In any situation there will be specific ways in which structural patterns have been understood, accommodated and the resultant projects expressed in formal institutions and ordinary life. These political processes are intensely contested and the out turn, a particular balance of contending forces,[2] will find a broad ideological expression in the sets of ideas carried in routine practice. These ideas actively mediate the relationship of structures and agent groups. The ways in which we read the world are part of the business of making that world; ideology is shaped by given circumstances and in turn shapes them. So ideological schemes do not simply endure, rather they are maintained.[3]

The historical development experience of the UK polity (reviewed earlier) offers a partial insight into the ways in which the ruling elite has read unfolding structural change, formulated projects and effectively mobilized its population. Such mobilization entails a mix of force and argument, matters of the deployment of power and the task of securing legitimate authority. Thus far these issues have been treated diachronically, a tale told about history. A complementary synchronic analysis[4] considers the discourse of Britain/Britishness.[5] The notion of discourse is slippery but it allows access to the spread of available ideas about political-cultural identity that are carried in the formal institutions and routines of everyday life in the UK.[6] The focus will be on the official sphere, the sets of ideas promulgated by authority and the ways in which these ideas pervade and shape the wider social world.[7] As the structural location of the British elite has changed down through time, so have their ways reading, reacting and formulating projects, and so by extension have their political-cultural identities. However, we can posit an irreducible core, a mix of praxis and ideas – the discourse of 'Britain/Britishness'. Three lines of analysis are presented: (i) the great tradition – the official sphere of Britishness (and its popular expression); (ii) the acquiescent little tradition – the accommodations of the masses; and (iii) the sphere of popular political thinking – the active political ideas of the masses. These analyses offer a way of structuring the otherwise fleeting world of political ideas and actions, the better to grasp its logic.

Great tradition – official ideology and its informal extension

The British polity represents itself in a particular way; one might speak of a great tradition (or an official ideology). Audiences are invited to acquiesce in a particular vision of the polity. Domestic audiences have little choice but to accommodate these claims as best they can, others are more robust.[8] The set of ideas details and celebrates the historical particularity of the British[9] (it is a simple nationalist perspective). After Robert Redfield[10] one might speak of the polity's great tradition; that over-arching set of ideas expressed in particular concrete institutional arrangements and carried by the distinctive patterns of life of the elite that legitimate and order extant social patterns. In contrast (and in opposition[11]) to the various political projects derived from the modernist project,[12] the official ideology of Britain reads the legitimacy of current arrangements out of the history of its ruling group – in Weberian terms a curious retrospectively oriented collective charismatic authority.[13] The ideology celebrates continuity and evolution, duty and obedience and an accumulative ideal of prosperity.

The great tradition has a core set of ideas: (i) constitutional monarchy; (ii) parliamentary sovereignty; (iii) representative democracy; (iv) the rule of law; and (v) the rhetorical schema of 'civility' and 'liberal individualism'. These ideas have both official ideological expression and popular variants. The reactions to these elements vary across the population according to the usual sociological reasons (age, class, gender, ethnicity and so on), however, some are broadly negative (as to politicians, bureaucrats), and some are broadly positive (the monarchy, witness the reactions to the death of Princess Diana).

Constitutional monarchy, parliamentary sovereignty, representative democracy and the rule of law

The British polity is a constitutional monarchy. The head of state is the monarch and the apparatus of the state is nominally the servant of the monarch (members of parliament swear an oath of loyalty to the monarch). The current system is the contingent outcome of the political exchanges that have taken place over the centuries (in other words, the system has a history, not an evolution, reaching back to the English Civil War). The current pattern locates the constitutional monarch within the apparatus of the state. The monarchy has a restricted practical function[14] and an extensive ideological role.

The institution of the popular monarchy is the ideological capstone of the present pre-modern UK polity: in place of notions of liberty, democracy and equality, the subjects of the crown rest content with notions of fairness, decency, compromise and consensus. Tom Nairn argues that the monarchy provides a link between the mundane ordinary world and the mythic historical realm of the nation: '… [a] national-spiritual sphere associated with mass adulation, the past, the State, and familial morality'.[15] The monarchy plays a significant role in the ordinary political life of the community, and there is a large media industry devoted to it: books, journalism, television. The members of the 'royal family' are celebrity figures. Media reports are cast in terms of direct revelations; a spurious intimacy with the audience is secured. The institution and personalities fuse. They are offered as a central model of the polity. The royals are linked to the hierarchies of class; while they are presented in the media as a celebrity family they offer a deeper social model. In voice and social style they exemplify '... the structure of authority which defines other speech, conduct and people as … merely those of a class or region'.[16] And against those who would dismiss concern for culture understood in this way, Nairn argues that social manners express '… the deeper structures of society and state'.[17] In the British polity an archaic pattern of manners, that celebrates the division of society into elite and mass, also provides an over-arching mechanism of unity: 'The glamour of Royalty and the neurosis of "class" are two sides of the single coin of British backwardness.'[18]

The common view of the monarchy has varied throughout the modern period with periods of popularity and unpopularity. One might speak of the 'public monarchy'; the business of 'pageantry' is relatively new, invented to recover the popularity of the institution in the late nineteenth century.[19] This pattern of presentation and popular reception has its own history. In the years following the Second World War the presentation revolved around the person and family of Elizabeth Windsor. In more recent years the presentation/reception has taken the form of popular 'celebrity' (notably with the life and death of 'Princess Di'[20]).

It is clear that the political role of the monarchy is contingent: it has changed, and it can change again. The political function – as the sovereign nationstate gives way, as it seems to be doing, to post-nation-state forms of sovereignty – indeed must change, as must the role of the person of the sovereign and the ideological role. Walter Baghot spoke of the 'dignified part of the constitution', but dignity is not readily compatible with celebrity. The public role of celebrity is quite distinctive, marked by transience, irrationalism and commercialism. Celebrity is a product for the marketplace, and the marketplace always needs new products. After Nairn, one might say that whilst it is true that the British monarchy is a significant part of the ideological apparatus of the present political settlement it is also difficult to see how it will sustain that role (never mind the activities of its republican critics).

The second element of the great tradition is parliament, the putative location of political power and authority. The institution is variously presented as: (i) as central to the democratic system of Britain; (ii) rational and deliberative; and (iii) august and dignified. However, these claims are routinely criticized. A series of points can be made. First, the system is presented as the acme of democracy (the 'mother of parliaments'); but the system does not rest on popular sovereignty,[21] and has neither popular democracy[22] nor popular nationalism.[23] Then, second, the 'crown-in-parliament' is the source of the institution's authority; it is the constitutional settlement of the English revolution. It is a pre-democratic system in a double sense: it is not merely the vehicle of a southern commercial elite, but also a closed system with effective political power reserved for qualified insiders. Third, the system is routinely misdescribed, thus the institution of parliament is only one element of a much larger apparatus – the Whitehall/Westminster machine. In fact this machine is elaborate, with many centres of power, large resources and a distinct ethos – it is the 'permanent government'.[24] It has been a routine target for 'declinist' analyses, but these may well miss the point of the apparatus (and the fortuitous nature of the lamented British pre-eminence). It has never pretended to be a 'developmental state', it has been an engine serving only sections of the population.

The system embodies a particular view of the nature of institutionalized power and authority. The 'Westminster model' was shaped by one legacy of the nineteenth century, the doctrines of utilitarianism. David Marquand notes that: 'Jeremy Bentham ... thought that sovereignty was inherently unlimited ... [and] the Crown-in-Parliament must be absolutely and inalienably sovereign.'[25] The system requires that political power and authority be concentrated and protected against rival claims, which could be local or regional government, or supra-national bodies, or the law in the form of codified constitutions or bills of rights. The Westminster model as it was established was very flexible. It was out of reach of effective criticism (no institutionally located authority could over-ride it); it could order its own affairs as it saw fit and was thereby insulated from criticism, as there were no rules or ethics to which critics could appeal. It was free to conduct its business away from the public gaze. It could also absorb dissenting groups; indeed critics faced a strategic choice – to join the parliamentary game (cooption), or to endeavour to use extra-legal means to force change (trades unions or, as Nairn points out, riots and public disturbances).

The Westminster model suited the British nineteenth-century liberal market very well; a global trading system was run by the oligarchy most directly involved.[26] Reforms and redistributions of power were available on the basis of success within the global system; these were sought by subaltern and middle-class reformers and agreed by the elite (who thus kept control over these processes). The elite could give ground slowly, protecting their interests/project. The form of the state served this project, however, as the global system changed it produced new demands, new competitors, new industries and new strategies of governance. A particular source of competition, the start of Britain's slow, long-running relative decline, was found in the developmental states[27] of the mainland. In a prospective/diagnostic view, Marquand argues that the British state cannot function like the developmental states which are common on the mainland because these require explicit central direction and thus negotiated consent, or power sharing – a strategy not available to the British state.[28]

Marquand discusses a slow collapse in confidence in respect of the Westminster system in the wake of the shift of the state to an interventionist role in the 1960s, and following Middlemas[29] he suggests that a long-established system-preference for political stability (even at the cost of economic growth) was fatally disturbed. Once the mystique of the Westminster model was weakened, as it was by the failure to secure such economic growth, it rather tended to dissolve away, leaving a strong state system capped by a 'club government' whose authority is waning. One might add, in regard to parliament, that the institution's authority has continued to slide. Two lines of argument might be mentioned: the analysts of the nature of Thatcherism, where the state was put to work to carry the political project of neo-liberalism (parliament was merely a vehicle for the promulgation of a project envisioned elsewhere); and the analysts of globalization/postmodernity, where the power of the global knowledge-based marketplace escapes the power of the state and the scrutiny of parliament. In both cases the rational deliberative function seems to have been eclipsed.

In the sphere of informal political discourse the idea of parliamentary sovereignty is acknowledged in a pragmatic or cynical fashion. The British parliament is seen as a centre of power but is often simultaneously seen as ridiculous. It is a familiar view, but it is difficult to read it as the opinion of those with power and access, as scorn is usually the position of those without either. The following views might be cited: (i) it is a centre of power but it is treated as theatre, a spectacle, to be viewed from the outside, a place that belongs not to 'us' but definitely to 'them'; (ii) it is often taken as ridiculous, as a bear-garden, undisciplined, a source of endless amusement for newspaper commentators operating in facetious, impudent or irreverent mode (a familiar style in Britain from the eighteenth century onwards – Gilray to Scarfe – and again belonging to the relatively powerless); and (iii) it is also a source of pride with the British model presented as the mother of parliaments. Relatedly it is seen as remote: (i) if struggling with authority one can write to one's MP; (ii) remoteness is conducive to gesture-politics, either as supplicant as immediately above, or as protester; (iii) groups often lobby parliament in ritual visits to the Palace of Westminster; and (iv) once every five years (or so) in a great collective ritual one can take a short stub of pencil and mark a cross on a ballot form. Overall parliament is little discussed. There is no routine linkage of citizenry and parliament. The system neither needs nor solicits the involvement of the population.

The third element of the great tradition is democracy. The notion of 'democracy' has been widely discussed,[30] and can be unpacked in a variety of ways.[31] One familiar distinction is representative versus participatory models. The former can be associated with the traditions of liberalism (usually tagged 'liberal democracy'[32]), the latter with republican democracy. The political system of Britain has developed a version of 'representative democracy'. The institutions are the lineal descendants of the settlement of the English civil war. A distinctive view of the political collectivity is offered: (i) citizens are seen as in need of protection from governors, and each other, thus the role of the state is to protect individual persons and their property; (ii) regular elections enable the individual sovereign people – all with particular interests – to aggregate their concerns and choose from competing party groupings their representatives in the parliament; (iii) the state machine thereafter makes the decisions but it is constructed with a separation of powers to block any concentration or abuse of power; (iv) the system affirms a notion of the rule of law, formal and public statements of agreed rules, within which individuals can act; and (v) the sphere of state action is circumscribed, with state and civil society in principle separate.[33]

The British political system is quite distinctive. Not only is it not democratic, it does not fulfil the more limited criteria of a liberal political system: (i) there is no separation of powers as the judiciary are neither independent nor apolitical; (ii) the executive commands extensive and unsupervised power; and (iii) parliament, supposedly the key and core of the system, is unrepresentative, the electoral mechanism generates a systematically distorted result (a situation known and accepted by the main parties for decades) and it is the routine vehicle for the promulgation of sectional interest and bought lobby representation.

In ordinary discourse it seems that representative democracy is acknowledged, but it is an abstract theorem to be repeated if asked. It is rarely debated. The participation of the mass of the population is reduced to petitioning, protesting and participating in what Max Weber called 'plebiscitary democracy' via periodic elections. Ralph Miliband[34] argues that the British political system, as it has developed since the Second Reform Act of 1867, is not democratic, rather it is 'capitalist-democratic'. The primary role of the system is one of demobilization, and parliament plays a key ideological role in that it enshrines the electoral principle and thereby legitimises the government and political system. All official life comes to centre on parliament and all extra-parliamentary political activity is de-legitimated. Miliband argues that the eighteenth- and nineteenth-century ideal of modernity, with its political project of 'formal and substantive democracy' has been deflected by a distinctly reactionary and restrictedly democratic grouping, and the ideal now finds limited restricted institutional expression in the contemporary parliamentary system. The focus of the system is on the containment of non-elite pressures, and the ideology of parliamentary democracy serves both to disguise fundamental patterns of power and to legitimate the system.

The fourth core idea of the great tradition is the ideal of the rule of law. It is crucial to modern ideas of politics, liberal and democratic.[35] It is also crucial to English conservatism.[36] Once again, the British polity routinely claims pre-eminence for its own system – others are judged against it and often found wanting. Ewing and Gearty[37] suggest that the official British reading of these matters is seriously awry, and whilst noting that up until as recently as the 1960s there was confidence in the system, they identify a subsequent rapid collapse. Reviewing in brief fashion the judicial system of the UK they point to a slow decline in the judicial independence established in the late seventeenth century as the crown and parliament lost power to the executive and note that the 'lack of any real constraints on the executive branch has led to a crisis of over-governability'.[38] In Britain governments can make law with little effective check (unlike either the USA or mainland Europe). Ewing and Gearty argue that civil rights are not positively defined – the liberty of the individual cannot be positively asserted – it is merely a residual, the sphere of that which is not prohibited.[39] The authors conclude that 'civil liberties in Britain are in a state of crisis',[40] adding that the 'major source of the problem ... is a political system which has allowed the concentration of power in the hands of the executive',[41] and rejecting as palliatives those solutions which call for more parliamentary scrutiny, or a human rights commission, or a bill of rights, they insist that what is needed is 'major surgery to the body politic'[42] in order, in effect, to democratize the system.[43]

In the common informal political discourse of the UK the idea of the rule of law is accepted and barely discussed: the official ideology finds informal articulation in phrases such as 'there ought to be a law against it', 'you must obey the law of the land', 'it is the law'. However, as noted, lately there has been some scepticism about the extent to which extant institutions and post-holders do manage to secure the rule of law in practice: the police are doubted, with questions about their efficiency and even-handedness; lawyers are viewed with unease; the judiciary are now seen as occasionally or routinely 'out of touch'; and the law itself is doubted thus, 'one law for the rich and one for the poor'.[44] But all this scepticism seems to be inchoate. There is no articulated view in regard to this issue, or what if anything should be done about it.

Civility and individualism

The political system is essentially oligarchic. The official ideology offers a hodge-podge of celebration and justification through its core ideas, which have both official expression (institutions, personnel, official understandings etc.) and unofficial expression in the routine and diverse reactions and understandings of the population. These core notions of the official ideology of the British polity in their informal extension come together, it might be argued, in the uneasily intertwined ideas of civility (acknowledging the permissive function of the apparatus – it provides order (it claims)) and individualism (acknowledging the goals to which the apparatus is nominally directed – it enables people to pursue their own needs and wants (it claims)).

In regard to the notion of civility, the characteristics might be listed as: (i) reasonableness, the preference for understated political and social life (withdrawal); (ii) decency, the preference for acquiescence in received common opinion and action (quietness); and (iii) moderation, the preference for limited appetites, demands and consumption (gentility). One might argue that this set is not unattractive – at the very least the people are 'amiable' – but it is in the nature of a consolation; the social style of civility in place of the political rights and duties of citizenship.

In regard to the notion of individualism, the characteristics may be listed as follows: (i) individualism, following the informal expression of political liberalism, affirms the priority of the person (rather than networks of relationships); (ii) informally, there are only individuals, and thereafter those institutional arrangements that individuals freely contract; and (iii) finally there is the generality of individuals, the mass. In this scheme, each individual confronts the social world thus understood as a realm of potential gain and loss, with other individuals as potential allies or opponents. One might argue – as opposed to the comments added to the discussion of civility – that this is a rather unattractive package. One might also argue that it cannot be widely translated into practice. The individualism celebrated by liberalism is essentially asocial; it cannot be attained. A gesture towards the spurious goal can be found in the expressive consumption of the media/marketplace.[45] One might argue that a natural sociability reasserts itself in everyday routine, but it is unacknowledged and subject to the media-carried celebration of individualism. It is not directly embraced, therefore not well developed. The notion of community appears only as an approved realm of self-help and ritual display (community centres, local sports teams, village fetes, and so on).

It is clear that there are multiple tensions in this pairing of civility and individualism. Both act to misdirect; civility and individualism share a latent infantilism and neither work to foster a socially located citizenship. For the moment the political-cultural dilemmas are resolved via a routine habit of deference: individuals are free to defer and to conform. One can pursue one's individualism within restricted private spheres; one can choose from available schedules of consumption. The boundaries of socially acceptable behaviour are clearly drawn, and it is a conservative society (in this it may well be like many other societies). It is important to be clear that this familiar pattern of ideas cannot be assimilated to mainland notions of the public sphere, an area of routine citizen political engagement, and nor can it be assimilated to the notion of civil society, the sphere of private economic activity. British political culture is distinctive – neither democratic nor liberal – and the cultural models invoked are often archaic. Patrick Wright[46] has noted the English celebration of a mythic rural past: hierarchical, deferential and emphatically pre-modern. It is the core of a national past which threads itself through contemporary society and offers an approved strategy of understanding/deferring.

An acquiescent little tradition

The British political system acts to demobilize the majority of the population.[47] This is accomplished by a series of mechanisms: first, institutional/procedural measures; thereafter, propaganda, persuasion and finally force (the usual machineries of universe maintenance). An elite great tradition is promulgated – 'Britain/Britishness' – which invades the space that might otherwise be occupied by popular democratic national ideas.[48] The popular sphere is turned inwards, to the domestic realm and the attenuated resources of extant community: an acquiescent little tradition. The population is thereby intellectually confined, restricted to a contested sphere shaped by elite demands and popular practice: one might say that Britishness competes with Englishness. The resultant cultural field is tangled and malleable: Britishness and Englishness are available resources and their use inflected by class, location and mood.[49]

Separating out British and English is awkward; here we make only some first moves.

In the work of James Scott[50] the idea of a little tradition is contrasted with the great traditions of high culture, the sphere of the elite. It is the autonomous moral community of local peasant life, and it exists in an essentially unstable relationship with the over-arching culture of the elite. The local form-of-life affirms distinctive ethics of action based on local cultural resources. The little tradition also offers the moral resources for dissent, both articulated and informal (everyday resistance).[51] The little tradition has been a resource as local communities negotiate the centralizing demands of the modern world.[52]

The extent to which it might be possible to speak of autonomous little traditions is unclear. The economy and society are highly developed. In the Victorian era a series of great cities with power bases distinct from London developed. These centres of economic, political and cultural power were separated from the metropolitan centre and from core groups of the British ruling class; that is, they were industrial capitalists not commercial, landed or mercantile traders.[53] Yet these groupings have been in slow decline throughout the twentieth century; the world of distinct regional or clearly marked class identities moved to its close in the early post-war years.[54] One might more plausibly speak of a 'mass society'.[55] The mass media is pervasive. However, the population is not homogenous: identity is constructed around locale, network and memory. Place is therefore crucial. There are residual regional city identities (Newcastle, Liverpool, Leeds, Manchester, Birmingham); there are distinctive small towns (Cheltenham, Winchester, Norwich, Canterbury); there are areas of light-industrial suburbia (M42 or M40 corridors); there are areas remote from the metropolitan core (Northumberland, Cumbria, Cornwall). There is also the distinctive metropolitan core of London and the southeast. It is not clear whether these differences count as residuals (the remains of once distinct economic, social and cultural areas – the base for local 'heritage') or potentials (genuine differences which might in other circumstances (England relocated) count for more). The issue of 'difference' has been considered in recent years, and more obviously, there has been significant inward migration – there are now visible ethnic communities.[56] As political-cultural identity is the outcome of the play of contending class groups, that is, contested, it may be that there is movement. However the key is the still vigorous dominant ideology of Britain/Britishness and local autonomy is over-ridden.

So the great tradition contains the four core ideas of monarchy, democracy, parliament and law (all with informal extension), which have a further expression in the guise of the pairing of civility and individualism (as modified by post-war American hegemony); together these over-ride any local and autonomous cultural schemes and the ideas of the elite pervade society. The result is an acquiescent little tradition. An elaborate repertoire of ideas can be discovered which have formal expression in the worlds of arts and literature and informal expression in the practices and conversations of everyday life; and they can be approached around the paired concerns for order and expression, that is, civility/individualism.

The mass, law-and-order and the bureaucrats

The logic[57] of liberal-individualism, as it is routinely evidenced in the day-to-day discourse of ordinary people,[58] entails that adherents take the view that there are only individual persons. Autonomously arising needs/wants must be satisfied in a social world governed by private calculations of gain/loss and ordered by contract. There are islands of naturally given surety, most importantly the family (and beyond that the larger family of nation), but the wider social world is an intrinsically insecure sphere.

The liberal philosophical ideal of negative liberty[59] makes the social sphere problematical. Outside the safety of the family, it is either the realm of contractually agreed order (centred on the state), or the realm of aggregated individuals, the mass. The mass is pre-social, unstable and governed by imposed rules that secure the convenience of individuals. The social sphere is not taken to be routinely and extensively structured, so law and order have to be imposed. The sets of imposed rules can suffer breakdowns as law and order fails, and this can lead to mob violence and the rule of the mass. In eighteenth- and nineteenth-century political commentary fear of the mob figured quite strongly[60] and such anxiety is routinely reproduced. Over the post-Second World War period there has been a series of moral panics, mostly focusing on the young, which recorded various threats to the fabric of society:[61] teddy boys, mods and rockers, hippies, football hooligans, welfare queens and so on. A preoccupation with law and order follows: it is a logically necessary anxiety.

A related sphere of potential restriction is found in the state and its agents, the bureaucrats. In the tradition of liberalism freedom is understood as freedom from constraint and the sphere of the state necessarily appears as restriction. A minimum spread of measures may be required to secure order, the condition of the pursuit of private satisfactions, but these restrictions are both diminutions of pre-given negative liberties and read as likely to expand. This fear finds expression in common discourse – it is the threat of 'bureaucracy'.[62] The bureaucracy becomes the realm of state-occasioned unfreedom. The bureaucrats paradigmatically deploy rules, which are obscure, fussy and lack common sense,[63] and on the basis of these rules interfere in the lives of ordinary people. Such interference impinges upon privacy and is intrusive and meddlesome. The bureaucrats are also taken as invariably inefficient, deploying red tape to enforce their meddlesome rules and having their own internal agendas.[64] They are often remote, uncaring, unresponsive and rude. Overall what they typically do not provide is good service; they are not deferential; they do not know their places as (unfortunately necessary) servants of the public.[65]

In the discourse of liberalism politics is in essence contractual. A delicate tracery of individual contracts sustains the political sphere that centres on the minimum state, the rule-giver and guarantor of order. There is no idea of collective political life, or legitimate action outside the specified institutional arenas.[66] All political action must revolve around or take place within parliament.[67] The idea of citizenship, with its legitimate and encompassing public sphere centred on the republican democratic state, is not available in British political culture.[68]

The informal and the eccentric

Isaiah Berlin[69] has drawn a distinction between positive and negative liberty: the power to achieve objectives versus freedom from external constraint. Berlin argued for the latter; the key to liberal ideas. The preference for negative liberty finds expression in the celebration of the informal and the eccentric, and the key is the absence of constraint. The acknowledged social is dangerous because it is either the sphere of the state (seen as the formal realm of contractually agreed order, and therefore necessarily restrictive) or it is the realm of aggregated individuals, the mass (which is not read as an informal realm of freedom but as a threateningly pre-social sphere). Liberalism offers no organic community ('there is no such thing as society'); both state and mass imply constraint or unfreedom (de jure or de facto). Resistance is appropriate and can be found in the informal and the eccentric: the informal is a phenomenon of the small scale (informality in the mass makes it a threatening unreasoning primitive force (not ordered)); eccentricity can be found in the domestic sphere or community, such figures are exemplars of freedom from constraint, those whose particular rule-breaking affirms the generality of the rules.[70] The ethos of informality/eccentricity is a reassurance.

In UK culture the informal is widely stressed in celebrations of the unstructured, the unordered and the spontaneous. The following are examples: (i) the familiar ideals of countryside which celebrate the rural scene as one of small-scale intimate reassurance, familiar and dignified by long historical existence, and celebrated in the visual arts, thus John Constable or the many portrait painters of the rural squirearchy;[71] (ii) the art-form of garden-making with its careful recreations of what is seen as unstructured countryside, and the related celebration of Garden Cities;[72] (iii) the familiar celebrations of village life and architecture (and cottage gardens) where the organic growth and pre-modernity of these forms is stressed, the 'village' of common understanding is almost a refuge from the contemporary world;[73] and (iv) an anti-modernism in architecture and the preference for the styles of earlier years generally, from medieval through to nineteenth-century Victorian houses, forms of urban living which accumulated over time and which were not rationally designed or laid-out.[74] One might also note that affirmations of the unconstrained extend to replies offered in regard to patterns of behaviour prescribed by the state with (early) objections to seat-belt laws, anti-smoking regulations or recently formal ID card systems cast in terms of such rules infringing on personal liberty.

The ethos of informality can be given specific vehicles in the persons of eccentrics: (i) Francis Drake, a war hero who took time-off to finish a game of bowls before sailing into battle; (ii) Robin Hood, a man who lived in a forest and stole from the rich in order to give to the poor; (iii) Sherlock Holmes, a late Victorian (fictional[75]) hero, who lives alone, plays the violin, takes opium in a 10 per cent solution and invents the science of detection;(iv) various criminals such as the Kray twins and the Great Train Robbers; and lately (v) media personalities, thus Fred Dibnah (steeplejack), Keith Floyd (cook) or Patrick Moore (astronomer) etc. Indeed, the celebration of the eccentric can extend to whole episodes which are remembered in a glow of informality: thus the Second World War was won by brave individuals (the few); eccentric rag-a-muffins (Dad's army); the scruffy (air-force personnel); with the whole enterprise exemplified by the make-it-up-as-we-go-along informality of the little-boats at Dunkirk (itself an informal episode). In addition there was a black market, with spivs and scrounging. The humour was supplied by ITMA and later the Goons. Against this we can note that as the myth of informality was created the reality, now hardly detectable in common political discourse, was of a command economy in the UK, with the USA and USSR providing money, munitions and armies.

The free consumer, thinker and actor

Alasdair MacIntyre[76] points out that if an official ideology is to successfully fill the available political-cultural space, those to whom it is addressed must dwell within its reach. Its integrity must be protected. A series of defences are available: institutional/procedural (parliamentarism), intellectual (the claims of the great tradition) and political (critics can be marginalized). But the great tradition also contains lines of argument that act to block public criticism; these point to the sphere of private subjective experience. The idea of the autonomous individual finds expression in a variety of social contexts – one set of claims acts to buttress the general position by restricting the possibilities of critical examination. In place of the public examination and judgment of claims (against social/public criteria) there is a preference for private/subjective claims – one might point to two lines of misdirection and an area of further background support: the familiar realm of choice and opinion, and thereafter the available distinction 'objective/subjective', itself an element of the deep-seated cultural assumptions of empiricism.[77]

The liberal social world is a realm of individual choices. The liberal-individualist construes action as essentially private, originating within the sovereign autonomous person. In a social setting any action that is outward oriented must conform to the body of contracted rules. However this requirement is external/pragmatic, rather than internal/moral, and obedience to the body of contracted rules is a pragmatic requirement occasioned by machineries of rule-enforcement. In turn, exchanges with other persons are a matter of the egocentric pursuit of private ends within the framework of enforceable rules. Frederic Jameson[78] has pointed out that the media-ordered realm of mass consumption offers an exemplar of such choices – the individual self creates itself freely in the spread of choices made within the consumer marketplace.

The liberal social world is also a realm of subjective opinions: one is entitled to one's opinion. Any claims to fact are to be judged against reality, the sphere of objects can be measured/modelled – it is quantifiable – but most social, political, aesthetic and moral judgements admit only of opinions, mine, yours or the guy next door's. In claiming this entitlement to opinion, the self that utters the opinion is the ground of its validation: that I assert my opinion is enough to validate the claims I make.

In the background there is the philosophical tradition of empiricism: the ground of all substantive knowledge claims is sense-experience, the necessarily individual sensory exchange with the given material realm. The facts of the world impress themselves upon us. The material realm admits of objective description – natural science – but the social sphere is one of mere habits and social practices – a realm of essentially subjective opinion. The empiricist position is discredited in epistemology but an echo persists in the sphere of social/political claims. In the social/political sphere matters are either quantifiable (and thus taken as objective and, like rules, commanding of obedience) or they are unquantifiable (and thus taken as subjective, and command no obedience). The sphere of the social is analyzed after the style of the natural sciences – it generates the intellectual habits identified as positivism.

All these ideas carry over into the public realm. MacIntyre[79] has argued that liberalism has no ethics, and that with its celebration of individual sensibilities, disregard for expectations of action carried in tradition, and pragmatic accommodation to enforced formal rules, the statements in regard to ethics which it does offer can be characterized as the manipulative ethics of emotivism.[80] The social/political sphere becomes reduced to the chaos of spurious claims to bureaucratically ordered technical objective knowledge confronting a haze of personal, and personally validated, opinions. The marketplace is a sphere of aggregated subjective choices. The absence of a genuinely social sphere means that bureaucratic claims evade criticism, so too the patterns produced by the marketplace, and personal opinions are valid-as-asserted. Without the mechanisms of public dialogue the population is in practice demobilized, and the social/political sphere is thus rendered inoperative. The boundaries of the great tradition are protected.

Tradition and nostalgia

A further line of defence is available in the guise of tradition and nostalgia. A collective preference for sentimental recollection (convenient untruths) helps protect official truths from historical scholarship. Wright[81] has analyzed the way in which a sense of history permeates the routine of everyday life. Everyday life is the realm of routine; stories about a national past are the means whereby the routines of everyday life are linked with the ideas affirmed by the wider group with whom we identify, the nation. The national past comprises the stories we routinely assume about ourselves, rather than explicitly tell.

In everyday historical consciousness agents explain their world and selves via plausible tales, stories, narratives or histories. It is a part of the business of making sense. It is embedded in routine, the sphere of the taken-for-granted. It is also strongly implicated in the response of groups and individuals to change. In the modern world we note, in particular, the rise of natural science and the consequent disenchantment of the world. Wright argues that the national past is a story told to inhabitants to make things meaningful in the context of the flux and disenchantment of the modern world.

In Britain nostalgia plays an important role. Wright offers the examples of the model of the old established family with its below-stairs retainers, the use of craft models of manufacture and the invocation of old character types. All of these act to structure the present, as the inheritor and present expression of the national past. Wright mentions two particular styles the national past adopts: the use of auratic sites where history is taken as most particularly present; and remembered war which is taken as a sphere of non-routine when actions made a difference.

Wright points out that there is a clear political aspect to all this as the group to whom we are routinely enjoined to belong is the nation. We are offered a national past, an official history. The nation replaces other more local communities and it re-enchants the world. A story is woven which tells us who we are, where we came from and where we are going. Wright continues:

> In the end, however, we come back to history in a more familiar sense, for the national past is formed within the historical experience of its particular nation-state. Among the factors which have influenced the definition of Britain's national past, therefore, are the recent experience of economic and imperial decline, the persistence of imperialist forms of self-understanding, early depopulation of the countryside, the continuing tension between the 'nations' of Britain (Wales, Scotland and, most obviously, Ireland), the continued existence of the Crown and so much related residual ceremony, the extensive and 'planned' demolition and redevelopment of settled communities which has occurred since the Second World War, and the still living memory of a righteous war that 'we' won.[82]

The political issue for Wright lies in the question of whether or not

> everyday historical consciousness might be detached from its present articulation in the dominant symbolism of nation and drawn into different expressions of cultural and historical identity.[83]

Common-sense political thinking

The political-cultural project of Britain was a coherent response to the structural circumstances of the elite in the wake of the loss of the American colonies, conflicts with revolutionary France and domestic pressures for democratization. The project attained a pinnacle of success in the early years of the twentieth century, and responded effectively to subsequent challenges. It affirmed a core set of ideas: those key to the great tradition that in turn feed an acquiescent little tradition. The British polity has an extensively demobilized population, but this does not mean an absence or end to politics. The concerns emerge elsewhere. There is a realm of common-sense-level politics in which three spheres of activity can be identified: the established political system; the realm of the market; and the sphere of the welfare state.

Politics: the style of deference and protest

The master style of deference both disguises and justifies an essentially oligarchic polity. The broad population is demobilized and has no grasp of the levers of power. It has a political role equivalent to that of a spectator, essentially passive – one can defer or protest. The master style of deference both disguises such demobilization and justifies the tacit acknowledgement of the claimed superiority of the ruling group. They have access to the knowledge and information; they have skills to judiciously govern; they are responsible; they should be trusted, left to get on with the job, not disturbed.[84] The broad mass of the population is directed into an informal politics; rehearsing arguments from media sources, or appealing to common sense, or reporting subjective feelings. The public sphere is drained and subjective protest confronts bureaucratic rationality.

The recourse to media sources is in the first instance entirely rational. In a developed democracy one would look to the media as a vehicle for the constitution of public discourse. However, in Britain the media typically does not play such a facilitating role. Analyses of public-service broadcast media reveal an orientation towards the status quo.[85] Commercial media organizations are increasingly part of global communications corporations with interests in news, entertainment and specialist information services. The sphere of the commercial media has expanded enormously, and commercially proffered market-centred life-style overwhelms the local and democratic. The initially rational impulse of the citizen, to consult the media so as to inform him/herself, is now fraught with difficulties. Against this, invoking common-sense has an important role within the informal political sphere – it signals an attempt to run arguments, to invoke little traditions.[86] After Frank Parkin[87] we could speculate that sources of common-sense ideas might be these: the neighbourhood, with concrete tradition, routine experience, gossip; the locality/region, with folk traditions; the workplace, with the individual/collectivity nexus, critical and realistic thinking; and organizations (private, voluntary), with a semi-formal realm, semi-formal opinions, plus perhaps specific knowledge of specific areas. An appeal to common sense is weak: it may be a sphere of accumulated knowledge, but, necessarily, it is unexamined knowledge. Finally, an appeal to subjective feeling is weaker still – at this point attempts at argument, explicit or stored, are set aside in favour of a spontaneous sentimental reaction to events (e.g. vigilante groups acting against alleged paedophiles, patriotic rallying to the state on the eve of war or the mass response to the death of Princess Diana.

As we shift away from the local sphere knowledge becomes more remote. The state, by denying local power bases and concentrating power in the metropolitan centre, pressures the individual to shift their sources of critical reflection from concrete tradition, which is potentially oppositional, a little tradition,[88] to the official realm of Galbraithian 'institutional truths', and thence to the fantasy worlds of the mass media. The masses are thus demobilized. Zygmunt Bauman[89] thinks this has been achieved, but it is engineered and needs must be continually re-engineered.

Given the lack of institutional vehicles for a participatory democracy, political argument emerges elsewhere, in the activities of marginal groups, and as the cultural/political style of protest. In the UK power is reserved unto the elites and the population are better seen as a mass rather than a citizenry.[90] As a cultural/political style protest is therefore well developed; we have parades, marches, meetings, riots and so on. It is an informal folk-politics, usually futile, but occasionally having an impact. On protest generally, MacIntyre[91] suggests that the modern world has come to neglect democracy in favour of a confrontation between subjective protest and bureaucratic pseudo-rationality.

Market: consumption and individualism

Mike Featherstone[92] reports that it is possible to identify three broad approaches to the analysis of consumer culture: the critical, the elucidatory and the affirmative.

The first critical line traces back to the analyses of Georg Luckacs and the Frankfurt School, who presented notions of mass culture, the superstructural form of Fordist modes of production. In brief the new realms of popular culture and mass consumption were taken to be both degraded cultural forms (low culture) and demobilizing politically (paradigmatically Hollywood, the 'dream machine'). Activities which had been undertaken in private time according to private or local cultural norms, became part and parcel of the mass cultures of entertainment or leisure. One familiar line of objection to such analyses is noted by Featherstone, the suggestion of elitist disregard for the ways in which consumer cultural forms were read by those participating in this realm.

A second approach to such questions can be derived from cultural anthropology where the issue becomes the use of commodities as cultural markers, and where such markers announce social position, often class. In this analysis the use of commodities within society is accorded a more straightforwardly social scientific analysis: the use is complex and subtle, and thus presents itself as a realm for investigation. Featherstone cites the work of Pierre Bourdieu[93] and his notion of 'habitus' – the way in which the class position of an individual is inscribed in patterns of consumption, bearing, taste, speech and so on. In regard to the theorists of postmodernity it is here that they begin their efforts: the contemporary world comprises a profusion of commodities amenable to subtle and routinely reworked significance. Indeed, as Featherstone notes, the contemporary consumer world throws up its own experts, para-intellectuals, those who teach consumers how and what to consume.

Finally, there is a line of reflection which traces back to pre-modern notions of excess that might be found in the institution of carnival, or the fair (where rational market exchange and expressive consumption are important).

Such forms of activity come down to the present, it is suggested, in the form of departmental store or shopping mall. Again, it is suggested that the critics of the Frankfurt School have missed an important issue, that of popular culture.

None the less, however the materials of the commercial world are read into ordinary practice it remains the case that the sphere is driven by the logic of the marketplace. The sphere of consumption/individualism is presented – by the commercial world – as a sphere of choice. The rhetoric is simple – a banal logical slide from free choice to free self to freedom-in-general.[94] Both the activity and the idea of consumerism reward as they demobilize. They also displace political activity. The celebration of consumption/individualism feeds into the ordinary discourse of politics in the guise of the routine displacement of the political. Issues are read out of the political sphere and read into the market sphere: (i) green politics becomes green consumerism; (ii) freedom of action becomes freedom of consumption; (iii) freedom of thought and association becomes life-style choice; and (iv) scholarship as central to discourse-democracy becomes technical expert advice oriented to the policy agendas of the elite.

Welfare: entitlement and help

The ideology of the welfare state is firmly in place, and the institutions and ideas run through the UK social world. It is deeply and ambiguously politicized: mis-direction plus support.

The mis-direction revolves around the translation of politics into welfare. In Britain socialism has come to mean welfare payments and transfers. At the present time, the sphere of welfare is a large area of entitlements (bureaucratic rationality confronts claims to rights[95]). The provision of welfare is now a major sphere of state activity.

The support is evidenced in the deep commitment of the population to the welfare state. The institutions are read as 'ours', political and financial (tax/expenditures) territory won from the elite and in need of defence.

Contingent responses to unfolding change

Contemporary 'Englishness' exists as a distorted residue of pre-British days that is overlain and obscured by a disintegrating Britishness and routinely disfigured by contemporary market-dominated practices and messages. And it is clear that all of this would be of merely archaeological interest were it not for the impact of recent patterns of structural change within the regional and global systems. It is within this quite particular context that one can speculate about a diminution of the reach of the notion of Britishness and the simultaneous development of a European political space within which the notion of Englishness might find new expression. All these are matters to pursue, but first we can note the depth of the shock to the elite of recent changes and the price they must pay.

Notes

1 This chapter is derived from P. W. Preston 1994, *Europe, Democracy and the Dissolution of Britain*, Aldershot, Dartmouth, chapters 2, 3 and 4.
2 On the fluidity/contingency of social relations and collective projects, see N. Long (ed.) 1992, *Battlefields of Knowledge*, London, Routledge. The same approach is adopted in regard to the EU by Jeremy Richardson – see J. Richardson (ed.) 2001, *European Union: Power and Policy Making*, London, Routledge.
3 P. Berger and T. Luckmann 1966, *The Social Construction of Reality*, Harmondsworth, Penguin.
4 A strategy insisted upon by A. Easthope 1999, *Englishness and National Culture*, London, Routledge.
5 The discussion will advance matters theoretically (if only slightly). In the introduction Englishness/Britishness were distinguished in crude class terms – Britishness was presented as an elite-sponsored official ideology that served amongst other things to suppress subaltern agitation for democratic republican reform, implicitly, English. The notion of identity was refined a little in the second chapter where an ethnographic/biographical notion of identity was presented (locale, network and memory). It was a species of substantive treatment stressing the rich detail of lived experience (illustrated with the help of Richard Hoggart). Political-cultural identity is one more layer of understanding. The notion of discourse opens up further areas of reflection: discourse-as-talk (or signal), the ways in which identity is marked in official/ordinary life; and discourse-as-narrative (Whitebrook 2001), the ways in which identity involves story-telling (a process which can be considered via its own technical requirements – narrator, voice and point-of-view).
6 Maureen Whitebrook 2001, *Identity, Narrative and Politics*, London, Routledge.
7 The more active responses of subaltern groups – aspiration, obedience, resistance or whatever – are not pursued. The issue of the resources available to challenge the dominant version of Britain/Britishness remains open: one could point to the wealth of ideas available within established institutions – those supported by the population (NHS, BBC, the school system or the emergency services); the energetic public sphere – media, organizations, clubs, societies and so on; the increasing ethnic diversity of the population – the resources of a host of little traditions; and also the increasing importance of those networks which route into mainland Europe, offering a further source of ideas. All these ideas are available

should contemporary patterns of structural change create new political spaces. Or, to put it another way, should the British oligarchy lose control, a popular national identity might look rather different to the received schemes of Britain/Britishness.

8 Thus, for example, in the autumn of 2002 and spring of 2003 – the period of the Iraq war – I was living and working outside the UK, in Germany and Hong Kong. There the UK elite/government was represented in the various public media as a mix of cliché (stereotypical Atlanticism) and sad farce (poodles).

9 M. Marcusen *et al.* 2001, 'Constructing Europe: The Evolution of Nation State Identities', in T. Christiansen *et al.* (eds) *The Social Construction of Europe*, London, Sage. Marcusen *et al.* speak of the myth of 'Anglo-Saxon exceptionalism' underpinning the British state and elite projects (p. 119).

10 R. Redfield cited in J. C. Scott 1977, 'Protest and Profanation: Agrarian Revolt and the Little Tradition', *Theory and Society* 4.1/4.2.

11 The early nineteenth-century Chartist aspiration to democracy was suppressed.

12 The core is the modernist project, the enlightenment's celebration of reason, but of course there are alternatives. The debate with the modernist project was extensive. At a general level once can cite nineteenth-century romanticism or the traditionalism/communitarianism of the (Catholic) Church. In terms of political programmes, critical reactions to modernity were voiced by conservatives, liberals and a variety of smaller groups.

13 The UK media routinely offer their audiences (the masses) opportunities to admire the elite with pieces on the royal family, country homes and gardens and various sporting events such as the boat race, Royal Ascot or Henley etc.

14 J. Paxman 1990, *Friends in High Places*, Harmondsworth, Penguin, discusses the monarchy as a constituent part of 'the establishment'.

15 T. Nairn 1988, *The Enchanted Glass*, London, Radius, p. 27.

16 Nairn 1988, p. 61.

17 Nairn 1988, p. 62.

18 Nairn 1988.

19 D. Canadine 1983, 'The Context, Performance and Meaning of Ritual: The British Monarchy and "the Invention of Tradition"', in E. Hobsbawm and T. Ranger (eds) 1983, *The Invention of Tradition*, Cambridge, Canto.

20 R. Colls (2002, *Identities of England*, Oxford University Press) suggests that the mass grief attached to Diana's death indicated a popular view of the Windsors – Diana was preferred (p. 378).

21 The crown-in-parliament is the seat of sovereignty – in another context, one of the first moves of the American occupation authorities in Japan in 1945 was to remove sovereignty from the emperor and lodge it with the people.

22 P. Calvocoressi 1997, *Fall Out: World War II and the Shaping of Postwar Europe*, London, Longman p. 135.

23 B. Anderson 1983, *Imagined Communities*, London, Verso.

24 Peter Hennessey 1989, *Whitehall*, New York, Free Press.

25 D. Marquand 1988, *The Unprincipled Society*, London, Fontana, p. 9.

26 Marquand 1988.

27 The notion of 'developmental states' is usually deployed in analyses of the success of East Asia in the years following the end of the Pacific War – its root is nineteenth-century Germany – but the term is apposite for an elite presiding over a national project. In the European context the more familiar term would be corporatism.

28 Marquand's book was both analysis and prescription and in it he made a rational and attractive case for the reform of the UK – roughly, Europeanization. This is fine, but the flaw so resides in its volutarism. It is an appeal to the state to take a new direction – but why should it? It serves its masters – the elite – well. Ruling-class success is perfectly compatible with national decline (Preston 1994), so the question should be – where is the force which will oblige the ruling class to shift their ground and open up new political spaces/opportunities?

29 K. Middlemas 1979, *The Politics of Industrial Society*, London, Andre Deutsche.

30 See D. Held 1987, *Models of Democracy*, Cambridge, Polity.

31 Philosophically – an ethic of collective social interaction to unpack; theoretically – a model of possible political practice to unpack; practically – a set of actual systems reporting themselves as democratic. All these approaches are amenable to analysis by political science. The whole business can be read historically: which states, what sort of claims and what sort of current judgements/lessons.

32 On these confusions see C. B. Macpherson 1973, *Democratic Theory: Essays in Retrieval*, Oxford University Press. While 'liberalism' and 'democracy' both make sense, 'liberal-democracy' does not; Macpherson argues that liberalism has had its day and liberal-democracy is a failed effort at delay.

33 One might mention two proponents of liberalism, Isaiah Berlin and Karl Popper. The important liberal/left theorist has been John Rawls. One might mention as critics, C. B. Macpherson, R. Plant and A. MacIntyre, as well as J. Habermas who presents the idea of discourse democracy. There is also a distinctive tradition of conservative thought, recently celebrated by Roger Scruton.

34 R. Miliband 1982, *Capitalist Democracy in Britain*, Oxford University Press.

35 Macpherson 1973.

36 R. Scruton 2001, *England: An Elegy*, London, Pimlico.

37 K. D. Ewing and C. A. Gearty 1990, *Freedom under Thatcher*, Oxford University Press.

38 Ewing and Gearty 1990, p. 6.

39 Ewing and Gearty 1990, p. 9.

40 Ewing and Gearty 1990, p. 255.

41 Ewing and Gearty 1990.

42 Ewing and Gearty 1990, p. 275.

43 The European aspects of law are pursued in Jo Shaw 2001, 'Postnational Constitutionalism in the European Union' in Christiansen *et al.* (eds). She argues that the law is one are where integration processes play out, and the outcome of debates/discourses cannot be specified.

44 For popular opinion see A. Sampson 2004, *Who Runs this Place? The Anatomy of Britain in the 21st Century*, London, John Murray, chapter 13.

45 F. Jameson 1991, *Postmodernism: Or the Cultural Logic of Late Capitalism*, London, Verso.

46 P. Wright 1985, *On Living in an Old Country*, London, Verso.

47 Crucially, the state machine is a part of the oligarchic system – centralized, secretive and largely unaccountable. The parliament is weak, the electoral system crooked, the parties largely interchangeable. On this and the diminution of the public sphere see D. Marquand 2004, *New Statesman* 19 January. See also D. Marquand 2004, *Decline of the Public*, Cambridge, Polity.

48 A point derived from Anderson 1983. The popular democratic national identities are exemplified by America and France while the elite top-down constructs, responses to the demands of democratic forces, are evidenced in Britain, Czarist Russia and India.

49 At the outset I suggested Britain was a top-down official identity and Englishness was the continuing residues of popular ideas, by implication, subaltern (as the elite bought into Britishness?). But this is too simple. The ideas of British/English need to be more precisely sketched and located; what do the ideas mean; what are their institutional carriers; and who are the people who buy into the idenities on offer. So here one way of beginning the task with some distinctions: class (obvious); location (the elite do not all live in/around the home counties they are present across the UK); and thereafter mood/voice (if Britishness is top down, the mood/voice is patrician/commanding (?), and if Englishness is subaltern then mood/voice are different – reserved/decent (?). Overall what we are looking for is a marginalized set of ideas about identity and we also ask what is their present expression.

50 Scott 1977.

51 J. C. Scott 1985, *Weapons of the Weak*, Yale University Press.

52 P. Worsely 1984, *The Three Worlds: Culture and World Development*, London, Weidenfeld.

53 Nairn 1988.

54 See R. Kee 1984, *1939 The World We Left Behind*, London, Weidenfeld; or R. Williams 1961, *The Long Revolution*, London, Chatto.

55 A. Kornhauser 1958, *The Politics of Mass Society*, London, Collier.

56 Yasmin Alibai-Brown 2000, *Who Do We Think We Are: Imagining the New Britain*, Harmondsworth, Penguin.

57 My formal intellectual sources here are Held 1987, Macpherson 1973 and A. MacIntyre 1985, *After Virtue: A Study in Moral Theory*, London, Duckworth.

58 See S. Lukes (1973, *Individualism*, New York, Harper and Row), who shows how the notion works differently in different Western countries – I am looking at the Anglo-Saxon variant.

59 Berlin's distinction – positive/negative liberty.

60 S. Schama (2003, A *History of Britain III: The Fate of Empire 1776–2001*, London, BBC, pp. 35–51) details the role of mob violence in the nineteenth century – revolutionary and Church and king.

61 The classic sociological text being S. Cohen 1972, *Folk Devils and Moral Panics*, Oxford, Martin Robertson.

62 See M. Albrow 1970, *Bureaucracy*, London, Macmillan, chapter 7.
63 D. Warwick 1974, *Bureaucracy*, London, Longman, chapter 6.
64 A recent theme of the New Right, see Marquand 1988.
65 Revealingly, the business of 'service' became one of John Major's early themes as prime pinister, and formed the core of his Citizens Charter.
66 The notion of civil society, the collection of private organizations, is familiar within liberal discourse. It is picked up by Jurgen Habermas and read as anticipatory of democracy. But its status is unclear with private contracts, private goals, sectional interests and lobby groups. It is read as vigorous. It has to be acknowledged, regulated, coopted and occasionally invoked, but power and control are reserved for the centre – the machinery of the state.
67 Miliband 1982.
68 Thus it impossible for the British state to constitute itself as a developmental state of the type familiar on the mainland – Marquand's point.
69 I. Berlin 1969, *Four Essays on Liberty*, Oxford University Press.
70 P. Langford 2000, *Englishness Identified: Manners and Character 1650–1850*, Oxford University Press, chapter 6.
71 J. Wolf 1981, *The Social Production of Art*, London, Macmillan.
72 J. Ryder and H. Silver 1985, *Modern English Society*, London, Methuen.
73 P. Jennings 1986, *The Living Village*, London, Hodder.
74 Wright 1985.
75 Arthur Conan Doyle was Irish.
76 A. MacIntyre 1962, 'A Mistake about Causality in Social Science', in P. Laslett and W. G. Runciman (eds), *Philosophy, Politics and Society, Series 2*, Oxford, Blackwell.
77 An idea pursued by Easthope 1999.
78 Jameson 1991.
79 MacIntyre 1985
80 B. Williams 1972, *Morality*, Harmondsworth, Penguin.
81 Wright 1985, pp. 6–7.
82 Wright 1985, p. 25.
83 Wright 1985 p. 26.
84 For an explicit expression of the technical variant of this see E. Gellner 1964, *Thought and Change*, London, Weidenfeld.
85 See for example the work of the Glasgow Media Group.
86 A territory also invaded by the elite. George Orwell's appeal to common sense, his trope of the plain man bumping into experience, worked to co-opt local tradition to the views of the establishment. See B. Crick 1980, *George Orwell; A Life*, Harmondsworth, Penguin.
87 F. Parkin 1972, *Class Inequality and Political Order*, London, Paladin.
88 On 'little tradition' see Scott 1977.
89 Z. Bauman 1988, *Freedom*, Open University Press.
90 One recalls Kornhauser 1958 from another period of conservatism.
91 MacIntyre 1985.
92 M. Featherstone 1991, *Consumer Capitalism and Postmodernism*, London, Sage.

93 For a review see D. Robbins 1991, *The Work of Pierre Bourdieu*, Open University Press.

94 Bauman 1988 points out that freedom is understood as life-style – see also Jameson 1991 on the seductive power of the media-carried marketplace.

95 MacIntyre 1985.

96 Nairn 1988.

6 Shock and price

The Second World War was a catastrophe for the British imperial political project. The political elite sought to sustain their international position: first, in terms of a central location within three intersecting spheres (the USA, the Commonwealth and EFTA);[1] later in terms of a bridge between Europe and America. This involved political and military dependence on the USA (decently veiled, with consolations plus cold war), coupled with an opportunistic economic stance towards Europe (understood as a free-trade area and global trader, EFTA II).[2] The sphere of policy has been appropriately ordered: towards America, subaltern obedience; towards European partners concerned with integration, dilute and delay.[3] This version of the established political-cultural project of Britain/Britishness was successful over years of the cold war. In the confines of block-time[4] the strategy made sense: block-time ensured American political presence and hegemony in Europe (thereby making space for British elite political status aspirations); it entailed a European military subordination to the USA (limiting European pretensions to independent action[5]); and it established a limit to European aspirations towards unification (and inevitable independence from the USA). The British political elite could treat Europe as a comfortable non-threatening economic sphere, safely lodged within the political ambit of the USA.

The events of 1989/91 overturned these settled arrangements. The end of block-time meant that the USA no longer had a rationale for its presence in Europe, and Europeans had no reason to curb aspirations to greater unity. An interregnum followed,[6] and productive debate began in the wake of the 2003 Iraq war.[7] The end of block-time also meant an end to the circumstances that sustained the British stance. As structural circumstances change, a response is inevitable (part necessary,[8] part chosen[9]). The episode of the Iraq war revealed the instincts of the oligarchy but circumstances will oblige an eventual accommodation with the project of Europe.

The shock of the new

As change swept through Europe, the British political elite had to deal with what Robert Hughes[10] in another context called the 'shock of the new'. A series of phases can be identified: (i) the events of 1989/91, with the autumn revolutions in Eastern Europe, the Soviet consent to these changes and the later collapse of the USSR; (ii) the 1989/91–2003 interregnum,[11] as confusion followed the end of the familiar post-war settlement; and (iii) fresh debate about the post-cold-war world initiated by the 2003 Iraq war.

The impacts of the events of 1989/91

The historical development trajectory of any polity will reveal a series of phases, interspersed with breaks as ruling elites read and react to changing structural circumstances. Any relatively enduring phase will find expression in a settled form-of-life. The shift from one phase to another will be unsettling.[12] The origins of change can be various. They might be endogenously generated, as with technological advance or new social groups or civil strife/ breakdown. They might also be exogenously occasioned, as with radical changes in enfolding structural circumstances. Any episode of rapid change is likely to be uncomfortable as the end points are not known. Any routes to the future must be designed with the intellectual machineries to hand: political ideals and experience, habits of policy-making and the resources of social scientific traditions. The events of 1989/91 marked the start of structural changes that were radically to remake the contexts within which British elite identities and actions were formed.

An anticipation of possible change came with President Gorbachev's perestroika and rapprochement with the West. The key events in phase one were the changes in Eastern Europe, where what had been anticipated in Poland with the rise of Solidarity and in Czechoslovakia with Charter 77[13] now unfolded throughout the Eastern block.[14] In June 1989 a non-communist coalition took power in Warsaw. In November of that year, after a summer of demonstrations, the government of the German Democratic Republic opened the Berlin Wall and resigned shortly afterwards, giving way to new political groups. In the autumn of that same year similar changes had taken place in Hungary, Czechoslovakia and Bulgaria. This sequence culminated in the Christmas revolution in Romania.[15] These events marked the end of the Soviet block in Eastern Europe; the Soviet Union acquiesced and withdrew its troops. There were cautious discussions about the new situation, regarding, for example, security architectures.[16] In 1991 a failed coup attempt saw President Gorbachev replaced by Boris Yeltsin, who proceeded to dismantle the USSR, re-inventing a pre-Soviet Russia, an unstable Commonwealth of Independent States (CIS) and seeing a series of territories in the Baltic and in Central Asia form new nation-states.

The first responses to change in Soviet block were informed by the idea-sets of block-time. The changes were welcomed as victory in the cold war and there were a series of Washington consensus-inspired reforms. These produced confusion. Yeltsin's Russia rapidly achieved the status of a Third-World country, with chaotic market reforms, gangsterism, cronyism and – evidence of the desperate plight of ordinary Russians – falling life expectancy rates. The changes were dramatic, and disentangling the consequences was difficult. The changes also had direct implications for the ways in which political actors in Europe, the USA and the UK understood themselves. After the initial celebrations it was clear that key elements in familiar self-understandings had been called into question: the post-war settlement, the rhetoric of cold war and the continued utility of the available consolations of the British elite.

It can be recalled that the end of the European civil war saw the continent occupied and divided – in the East, state socialism, in the West, the liberal trading sphere of the Bretton Woods system. The crucial over-arching intellectual and political framework within which American and Western European thinkers understood themselves was the US-sponsored cold-war political project of the free West. The rhetoric of cold-war competition was an expression of sets of ideas embedded within the institutions and practices of the post-war settlement – all the apparatus (intellectual, institutional, political and popular) which sustained the American-sponsored sphere of market economies, liberal-democracies and open liberal trading networks. The USSR offered an alternate vision of an ongoing route to the modern world which, in the wake of the depression of the 1930s and the fascist-inspired catastrophe of the Second World War, was attractive to many political groupings in Europe, Latin America and the colonized lands of Asia.[17] The USA's hostility to this vision should not be underestimated, nor should the accidental, even irresponsible, nature of the cold war.[18] However, once the apparatus was in place the competition was extensive and widely damaging domestically in the USA,[19] in Europe (where local communist and socialist traditions were deeply embedded within the political sphere[20]) and in the Third World.[21] The competition came to be a central defining element of the political thinking of the state-regimes of the West. Not merely a crucial aspect of the institutional machineries which vehicled the post-war settlement (thus, obviously, NATO), and not merely the coarse public expression of views held sincerely by some (the political right, and those theorists who lumped together the development history of the USSR with episodes of European fascism under the term 'totalitarian'[22]), but also a ritual truth,[23] a set of statements widely understood to be false but which none the less served the function of ordering discourse and social interaction.

The British political elite found consolations for their post-war loss of empire[24] within the American sphere. In the context of the post-war liberal reordering of the global system sponsored by the USA, the British acquiesced in their subsidiary status and consoled themselves with thoughts of Britain's 'Greece to America's Rome', the 'special relationship' and the 'English-speaking peoples'.

The implications of 1989/91 were profound. The rhetoric and reality of the post-war settlement were overthrown. The global system was not bi-polar; countries were pursuing a series of diverse routes to a modern world whose character continued to unfold.[26] The institutional machineries of the post-war settlement and the intellectual simplicities of the doctrines of the free West were clearly not adequate to the new situation. The events of the period undermined the British elite's post-war political-cultural project, its sense of its place and role in the world, and a decade of confusion followed. The EU has been important in domestic party politics: (i) Margaret Thatcher's (eventual) emphatic public rejection, itself rejected by her party, which thereafter slowly modulated into the semi-official position of the Eurosceptics with their aspiration to a recovered, US-centred status quo ante; (ii) John Major's pragmatic acknowledgement, which seemingly had no further substance save that of hoping that something might turn up; and (iii) Tony Blair's vacillation over the Union, with its rhetorical yes and practical no. It would be tedious to rehearse the details;[27] but we might note that these debates continue within the main political parties, amongst the Whitehall/Westminster village[28] and in the metropolitan media.[29] Such debate as there has been within the wider population merely reflects the debate of the political classes.[30]

Interregnum: 1989/91–2003

The 1990s saw wide debate about the nature of the post-cold war global system. There were a series of unconvincing calls for a continuation of American primacy (the end of history, clash of civilizations, globalization,[31] market-democracies, etc[32]). It was more productively argued that geo-strategy had given way to geo-economics. There were calls for greater international trade. The World Trade Organization (WTO) was established and there were major agreements on deregulation and liberalization. If the American reactions were celebratory, the European reaction was diverse; we can mention three countries and one institution: (i) in Germany the populations faced the task of reunification, largely successfully; (ii) in France the political elite responded with a renewed drive to unify Europe;[33] (iii) in the UK political elite reacted with dismay as their comprehensible congenial world dissolved; in order to reveal (iv) the newly prominent machineries of the European Union.

The implications for the British elite were ambiguous. The greater concern for the geo-economics of trade underscored their loss of political status; yet their economic situation was good. The withdrawal from the Exchange rate Mechanism (ERM),[34] inward foreign direct investment and privatization/liberalization stimulated domestic growth. However, German reunification put a nation of eighty million people at the heart of Europe and global trading nations could readily identify the centre of gravity of the European economy – it was not the UK. The domestic British calls for a 'third way' reflected the wider uncertainties. It was clear that any continuation of the post-war settlement was not possible – the demographic, economic and cultural weight of the EU was now apparent to all.[35] Yet in the Iraq crisis London reverted to Atlanticist type. Prime Minister Blair's relationship with Europe collapsed, as did his domestic popularity.

Spring 2003 – the start of debate

The events of the 2003 Iraq crisis are likely to be long debated. There are three areas of immediate interest. First, President Bush's foreign policy stance was unilateralist, expansionist and not averse to the use of violence. The assertion of a 'war against terrorism' was merely a call for a continuation of an unavailable cold-war leadership role. The enlightened self-interest that informed President Wilson's work on the League of Nations and President Roosevelt's with the United Nations was abandoned. It will take time to repair the damage. Secondly, the response of the members of the European Union was disunited: the political elites split (the masses did not, more or less uniformly opposing the American line) with one group affirming ideas of international law and the value of multi-lateral institutions whilst the other elected to support the American regime's choice for war. Again, the damage will take some time to repair. Thirdly, the response of the British elite was inept and revealing: they rallied to the support of the Americans. The extent of the damage to their position in Europe (and elsewhere) was readily apparent.[36] The crisis saw hostility between the USA, the British elite and the EU, and tensions between EU members. The deepening separation of interests between the USA and EU (a matter of long-term structural change) suggests that a return to the pre-invasion status quo ante is not possible.

The question that has haunted the British polity since the end of the cold war was presented once again: what is its place/role? The episode of the Iraq crisis revealed the instincts of the political elite to cleave to the USA, but this is unlikely to be sustainable over anything other than the short term.[37] The question is presented once again – just what is the anticipated trajectory of the Isles? As the crisis unfolded a series of preliminary positions emerged: (i) to confirm the project of Britain as an off shore part of the continuing American sphere; (ii) to embrace the lessons of political error and rally to the EU; and (iii) to pause for thought, assess the damage to British/EU relations, assess the new relationship of the EU/USA and then to look to the future. We might add to this (iv) the long-term view, which is that the British oligarchy must deal with Europe as they have nowhere else to go.

Anti-Europeans argued that Britain should remain part of the American sphere. There were two variants of this argument; the tacit and the explicit. The former suggested that the interlinkages of Britain and the USA and EU were deep and significant change would either be difficult or divisive (or both). A modest movement forwards was proposed; the EU should continue to develop slowly, NATO should remain the cornerstone of European defence and the issue of EU/USA relations should be set aside/fudged. This strategy had the advantage of being non-dramatic, but had the disadvantage of ignoring the key problem of the relationship of EU/USA. The explicit version argued the case directly: membership of the EU was a mistake; it should be rectified.

Pro-Europeans suggested that the British elite rally to the EU. It was argued, for example, that Prime Minister Blair was transformed by the invasion of Iraq – no longer weak and vacillating, rather strong, determined and able to over-rule doubters in his party, Whitehall and Westminster and call a referendum on the euro. It was an implausible line: polls recorded a hardening of popular opinion against the euro; key figures in his party remained hostile, so too those in Whitehall and Westminster. Blair looked weak, not strong (maybe even weaker than before, as the elite would have noted both that he 'got away with his policy' and that he neither commanded domestic support nor significant influence in Washington).

A 'pause for thought' was also mooted. It was suggested that before any new lines of advance were determined, time should be taken to assess the damage to British/EU and EU/US relations. It was only following such reflection that it would be possible to look to the future. It was not entirely clear what this might mean. There seemed to be two strands of thinking: first, to remake trans-Atlantic relations; and second, for the EU to attend to its own interests (both prospectively on the basis that it is a good idea, and reactively as the American government has made its unilateralist views clear).

The final scenario, the long-term view, argued that the British oligarchy had nowhere else to go; structural circumstances would oblige an accommodation with the project of Europe. It was pointed out, on the one hand, that there were deepening linkages with the mainland, and on the other that American foreign policy thinking is realist; hence the US foreign policy community was likely to pay attention to the powerful mainland core of Europe, rather than a semi-detached periphery.

Events and opportunities

The episode of the interregnum has allowed crucial issues to drift; in particular the nature of the British political elite's relationship to the EU. But that period is now over. The events of the Iraq war have re-centered the issue of Europe. A series of formal and substantive conclusions are available.

Formally, generally:

(i) we are reminded of the scope, richness and relevance of the classical European tradition of social theorizing; and

(ii) we are reminded of the scope, richness and relevance of analysis which looks at the sets of ideas that structure the actions of various agents within the social world (discourses, ideologies, world views, etc).

Formally, more specifically:

(iii) it is clear that actions cannot be read-off (imputed) material interests, because self-understanding, that is, identity (the ways in which agents locate themselves and grasp the world), is the basis for action (this recalls the work of Max Weber and cuts against ideas of orthodox market-liberals and (if they still exist) mechanical Marxists); and

(iv) actions cannot be read-off formal 'rational actor' models, because self-understanding, that is identity, is the basis for action (hence, social scientific understanding requires ethnographic research – 'thick description' – otherwise we cannot grasp the terms in which actors act).

Theoretically, generally:

(v) the business of structural change and agent response is given, in the sense that it is routine in social life; and

(vi) if change involves re-inventing identities and political-cultural identities, then the price is likely to be high and the task likely to be difficult.

Substantively:

(vii) the period 1989/91–2003 undermined the British political elite's view of the world, its understanding of its place within it and any plausible route to the future informed thereby;

(viii) adjustment to the new structural pattern demands deep-seated changes;

(ix) presently, all that has been achieved is deepening confusion; and

(x) finally, the price the British must pay to (finally) 'join Europe' is to leave behind the idea of Britain/Britishness.

Shock and price

The shock of the events of 1989/91 generated a period of confusion within the old block-time territories of the West and early attempts at the repair and renewal of official ideologies were unpersuasive; matters drifted until the events of 2003. The depth of the conflicts between the USA and the EU, and amongst European partners themselves, ensured there could be no return to the status quo ante. A series of questions were openly presented: what was the nature of the project of the European Union; what was the relationship of the EU to the USA; and (as one relatively minor element of these issues) what was the future trajectory of the Isles. The EU project is central to European politics, it might experience confusion – it has before – but it is unlikely radically to change direction. As regards the British elite, one might speculate that the strategy of delay will continue with only grudging accommodation to the European project. However, as contact with the EU increases it becomes evermore clear that they have nowhere else to go.

Notes

1 D. Urwin 1997, *A Political History of Western Europe Since 1945*, London, Longman.
2 P. W. Preston 1994, *Europe, Democracy and the Dissolution of Britain*, Aldershot, Dartmouth.
3 Preston 1994.
4 A phrase I take from E. P. Thompson.
5 This was made clear as the Europeans withdrew from empire – in the case of the UK the signal event was Suez.
6 T. Judt 2002, 'The Past is Another Country: Myth and Memory in Post-war Europe', in J. W. Muller (ed.) *Memory and Power in Post War Europe*, Cambridge University Press.
7 In the debates about Europeans being from Venus and Americans from Mars the language is childish and the tone unhappy, but the issue is a real one – what is in process is the rebalancing of an important relationship. This issue is one outside the scope of this text.
8 P. Winch 1958 (2nd edn 1990), *The Idea of a Social Science and its Relation to Philosophy*, London, Routledge.
9 A. MacIntyre (1962, 'A Mistake about Causality in the Social Sciences', in P. Laslett and W. G. Runciman (eds), *Philosophy, Politics and Society, Series 2*, Oxford, Blackwell) argues thinking about ideology will let scholars see why one course of action was taken. In a similar vein J. C. D. Clark (2003, *Our Shadowed Present: Modernism, Postmodernism and History*, London, Atlantic Books) argues for a'counterfactual history', sensitive to choice and contingency.
10 Robert Hughes 1991, *The Shock of the New: Art and the Century of Change*, London, Thames and Hudson.

11 Judt 2002.

12 E. Gellner 1964, *Thought and Change*, London, Weidenfeld.

13 See T. Garton-Ash 1990, *We the People: The Revolution of 89*, London, Granta.

14 Urwin 1997, pp. 291–7.

15 The Balkan tragedy unfolded a little later. See M. Ignatieff 1994, *Blood and Belonging: Journeys into the New Nationalism*, London, Vintage; D. Rieff 1995, *Slaughterhouse: Bosnia and the Failure of the West*, London, Vintage; M. Glenny 1992, *The Fall of Yugoslavia: The Third Balkan War*, Harmondsworth, Penguin; B. Simms 2002, *Unfinest Hour: Britain and the Destruction of Bosnia*, Harmondsworth, Penguin.

16 S. Croft, J. Redmond, G. Wyn Rees and M. Webber 1999, *The Enlargement of Europe*, Manchester University Press.

17 M. Mazower 1998, *Dark Continent: Europe's Twentieth Century*, New York, Alfred Knopf.

18 G. Kolko 1968, *The Politics of War*, New York, Vintage. For the alternate mainstream position, see J. L. Gaddis 1997, *We Now Know: Rethinking Cold War History*, Oxford University Press.

19 For a nice example, see Lillian Hellman 1976, *Scoundrel Time*, New York, Little, Brown.

20 On anti-communist paranoia in the UK, see P. Hennessy 2003, *The Secret State: Whitehall and the Cold War*, Harmondsworth, Penguin.

21 Fred Halliday 1989, *Cold War, Third World*, London, Radius.

22 So that their own unsavoury allies could be conveniently termed/excused 'authoritarian' and their opposition to any left/nationalist 'totalitarian' elites thereby justified.

23 A. P. Cohen 1994, *Self-Consciousness*, London, Routledge.

24 The end of the European civil war, the Second World War, saw the British empire destroyed – what had been a globe-spanning empire in the autumn of 1939 was in ruins by late 1945. A rapid series of exercises of decolonization took place in Asia and a little later in Africa. The last empire endeavour of the British, the invasion of Suez in 1956, was vetoed by the USA.

25 C. Hitchins 1990, *Blood, Class and Nostagia: Anglo-American Ironies*, London, Vintage. A 2003 view is put by Roderick Braithwaite 2003, *Prospect* March/April.

26 Neither globalization theory nor its predecessor, convergence theory, are plausible – the model of the USA is local, not universal (S. Gudeman 1986, *Economics as Culture*, London, Routledge).

27 But see Hugo Young 1999, *This Blessed Plot: Britain and Europe from Churchill to Blair*, London, Macmillan.

28 In the sphere of public policy making, the Whitehall/Westminster village, we might speculate that the evidence suggests that the party political and media positions are repeated, that is, an overall unclarity coupled with a disposition to delay. (An unclarity maybe compounded by departmental differences treasury, cabinet office, foreign office and defence ministry – where departmental views will cut into the issue of Europe/Atlantic in different ways.)

29 In the sphere of the media it is important to note the crucial role of the Eurosceptic press whose commentaries have intermingled irrational nationalism, calculated mischief-making, routine irresponsibility and a pervasive bias towards the local concerns of elites in London. The implications of the end of block-time have been disregarded or denied, while a routine celebration of neo-liberalism, the Atlantic alliance and the model of the USA continue. The issue of Europe meanwhile has been dealt with in a simplistic, hostile and – for their goals – successful fashion.

30 In the sphere of the ordinary population it is important to note, first, that the official ideology and institutional make-up of the British state serves to demobilize the population. There is no popular debate on Europe, reactions are piecemeal, inchoate and *open* (evidenced in: (i) practical activity, thus business links to the European Union, travel within the Union, and interestingly, the increased use of the European Union as an 'available good example' (thus the price of cars, or the scope of welfare systems, or even the alleged absence of football hooligans); and (ii) opinion surveys).

31 The only interesting debate has surrounded 'globalization' – on this see P. Hirst and G. Thompson 1999 (2nd edn), *Globalization in Question*, Cambridge, Polity.

32 See F. Fukuyama 1992, *The End of History and the Last Man*, London, Hamish Hamilton; S. P. Huntington 1993, 'The Clash of Civilizations', *Foreign Affairs* 72/3; P. Bobbitt 2002, *The Shield of Achilles: War, Peace and the Course of History*, New York, Alfred Knopf.

33 L. Siedentop 2001, *Democracy in Europe*, London, Allen Lane.

34 September 1992 generated a significant devaluation and left the authorities with no economic policy – it destroyed the reputation for competence of the Tory party.

35 G. Barraclough (1964, *An Introduction to Contemporary History*, Harmondsworth, Penguin) argues that one of the reasons for the decline of the British Raj (and wider European/American colonial holdings in Asia) was simply demographic: there were a few hundred thousand Europeans/Americans; there were tens of millions of Asians.

36 Marquand 2003.

37 The EU works as an economic block and the USA has to deal with this. Another economic area exists in East Asia. The US military has bases around the globe (the basis for misleanding claims to pre-eminence). In this context, the British elite do not count.

7 First implications

British political discourse has been based on a number of key assumptions in recent years: first, US economic, military and political leadership in Europe (coupled with a particular subordinate role for the British, the special relationship); second, a continuing adherence to notions of sovereign statehood and national self-determination (and thus an 'independent' political line); and third, the routine affirmation of an official ideology of liberal-democracy, the notion of 'the West'. However, the circumstances that occasioned this familiar political discourse have now changed and one might expect agents to formulate new projects.

Early British debate in respect of Europe

The response of agent groups within the UK to the structural changes of the period 1989/91 were mixed. The response of the political elite was negative, in line with earlier thinking on Europe. The opening phase of political discourse within the UK in respect of the European Union saw a narrow and pragmatic focus upon short-term goals.[1] The discourse was national, economistic and unreflexive.[2] The sets of ideas expressed prior to 1989/91 persisted into the mid-1990s, and whilst the 1997 election of the New Labour government seemed to signal a change, little of significance happened. In 2003, as the Iraq war unfolded, the British elite reverted to Atlanticist type.[3]

A national discourse

The public discussion of the relationship of Britain with the European Union was cast in terms of gain/loss and problem/opportunity. This style of discourse reflects not merely the personalized nationalism of the British ruling groups, whereby 'we' are taken as of one mind, and set of interests, all of which must be necessarily, inevitably and obviously asserted against 'them'[4] – in this case the rest of the EU – but also the habit of reading the world through the intellectual frame of an 'informal liberalism'.[5] In respect of the former, the official ideology reads legitimacy out of the history of its ruling groups (oligarchy) – the legitimacy of the state derives from the history and record of that elite (not any explicit political philosophy, event – e.g. revolution – or appeal to the masses).[6] The political elite can affirm that the future can be discussed in terms of gain/loss. But when matters are contested (as is the nature and dynamic of the EU[7]), when debate is concerned with institutional structures and patterns of identity, not merely technical matters of the governance of a modern state lodged in an internationalized/regionalized global system, then the response is a strategic mis-reading of the EU as something that it is not. Instead, the British elite attempts to present it as something that it wishes that it were, an inter-governmental contractual arrangement. In the latter case we are dealing with liberalism-writ-large: thus as discrete individuals are taken to interact in pursuit of their autonomously arising needs and wants, so too are nation-states. The public political discourse of Britain dealt with the dynamics of change in the EU as if it were a matter of the interactions of discrete units. However, any analysis that looks to the dynamics of structures and agents reveals the pursuit by the British political elite of quite specific political projects. The nature of the elite's exchanges with their partners in the EU was a matter of concern to commentators over an extended period and there were a series of recurring themes: the preoccupation of the elite with the issue of sovereignty; the strong preference for intergovernmentalism; and the way in which British political agents seemed to misjudge their colleagues on the mainland.[8]

A long history of problems was identified. The overall shift in the post-Second World War period was characterized as being from isolation to semi-detachment.[9] A central concern was for sovereignty, which figured not only centrally in British political-institutional traditions but also took a talismanic role in public debate such that all politicians were obliged to affirm its central status. Analysts spoke of the unhelpful logic of adversarial politics, intra-party unease and the challenge to the ideological self-images of the parties as guardians of respectively the British national interest and the parliamentary route to socialism.[10]

It was argued[11] that whereas the initial group of members of the EU and most of the 1970s and 80s joiners[12] had clear political reasons for setting up or joining the organization, the British in contrast were late, reluctant and opportunistic members for whom participation was an expression of the failure of their independent line.[13] The government found it difficult to engage effectively with the EU as the disposition to semi-detachment permeated the thinking of political, administrative and economic elites.[14] One way in which this was routinely expressed was in the systematic preference showed by British governments for intergovernmental solutions over federal strategies.

The upshot was pragmatism towards the development of the EU, which jarred with the expectations of the mainland members. Integration at the level of the machinery of government went ahead smoothly, but it remained the case that the British government was reactive and nationalistic.[15] In particular, British governments misjudged the commitment of mainland elites to political integration.[16] It was noted that the mainland view of the UK was negative.[17]

Against the familiar national discourse it was clear by the early 1990s that structural change within the EU was well advanced. The analytical issues centred on the extent of integration, the depth of its reach within the political-economic structures of extant nation-states and the pace of integration. These were a series of analytical and political issues that were far removed from anything within the public political discourse of the UK. It is not clear, from the vantage point of the new millennium, how much has changed: elite discussion is restricted and popular discussion muted.

An economistic discourse

The independent political project of the British elite, pursued in the years following the Second World War, came to an end in the 1950s when the debacle of Suez, the unpersuasive record of EFTA and the burgeoning success of the European Economic Community (EEC) persuaded the government of the day to reorder its thinking and apply for membership of the EEC.[18] This was presented as prospectively joining a 'common market'. The EEC was seen as a source of trade in a global system that had become less hospitable to the Britain and a series of economic advantages was promised. While this view was clear under Prime Minister Harold Macmillan, the premiership of Edward Heath was different as membership was seen in broader terms.[19] However, subsequent governments reasserted the economic pragmatism:[20] it was evident in Thatcher's demands for rebates; it was evident in Major's claims for the pound against the mark; and it was evident in the New Labour government's treatment of the question of the euro – a matter subject to 'the five economic tests'.

The early years of UK membership coincided with the end of the post-Second World War 'long boom'. These years were dominated by issues related to the maladaption of the UK to the EEC: in particular, budget contributions and the Common Agricultural Policy (CAP).[21] The former issue, where UK trade patterns entailed onerous financial obligations to Brussels, caused considerable controversy. The difficulties with the CAP were occasioned by the differences between domestic state support for agriculture and the practices of the mainland. Once again the process of resolution was long and acrimonious and debate ran on until the Fontainebleau Summit of June 1984 when Mrs Thatcher secured a rebate on the UK contribution to the community budget. In complementary fashion, the areas where the UK was most effective were those which favoured directly its own interests, in particular the drive to the single market and the outward-directed aspects of EU organization where the UK pressed for open trade.[22]

Much debate in the UK prior to 1989/91 was cast in technical economic terms, for example to join or not to join the ERM. The revival of moves to integration in the 1980s – the Single European Act (SEA) – was presented as completing the single market.[23] After 1989/91 it was granted in public discourse that the issue of the EU went wider than the narrowly economic, but the acknowledgement was grudging. It was often cast in terms of sensible British resistance to the fanciful demands of excitable mainlanders. The rhetoric has included more explicit comparisons – mainland 'economic rigidities' versus 'UK flexibility', the neo-liberal economic policy legacy of Thatcherism. All these patterns of debate have continued. The recourse to the language of economics, as an escape from politics, has been a consistent feature of the British elite in their public pronouncements.[24]

Against the familiar mainstream public debate it can be asserted that an economistic discourse is unsatisfactory because it extracts economic linkages from the broad spread of human exchanges (economic, social, cultural and political), and casts analyses in terms of precisely specifiable matters of the functioning of automatic systems. In debate about the UK/EU it was assumed that it made sense to speak of economies in isolation. However, economies are always embedded in societies (thus they are not asocial technical machineries amenable to naturalistic analysis), and national economies have extensive linkages to trans-national global networks. Henk Overbeek[25] has argued that the post-Second World War period saw the establishment of American hegemony over an Atlantic economic sphere. In recent years, in a similar fashion, it has been argued that the global system is becoming more integrated (internationalization[26]) and regionalized (Europe, North America and East Asia). At present, in respect of the UK economy one might say that the reality is of extant trans-European regional structures that are in turn lodged within internationalized global structures.

An unreflexive discourse

The British elite's characterization of the EEC in liberal contractual terms, representing it as a matter of simple choices, was parochial; it neglected the wellsprings of the entire enterprise.[27] The experience of war was fundamental to the formation of the EEC.[28] An economic motive was available in the concern to formulate some sort of response to early post-war US pre-eminence. The EEC was successful: extensive economic linkages were established, the record of the members was impressive, and related to these, the period saw the growth of extensive networks of political cooperation – most obviously in the establishment of the organization itself. In the 1970s and 80s the dynamic of the EEC faltered and doubts emerged. The response was a renewed drive towards realizing the original goals of the organization, cast in terms of the completion of the single market. The programme envisaged dismantling frontier controls, bringing product specifications into line, establishing rules about government procurement programmes, simplifying and harmonizing tax regimes, and freeing capital movements – all of which implied a vast schedule of detailed reforms. The drive to complete the single market also implied a large spread of social, cultural and political change, some sort of commitment to some sort of EU unification.

By the late 1980s it was clear to some that the post-war settlement itself was in question.[29] A series of alternative patterns of change were sketched: (i) Atlantic reformism which looked to a continuation of NATOism with a larger European role; (ii) European reformism which looked to a post-NATO system involving European federal integration; (iii) European 'Gaullism' which involved a post-NATO European nationalism; and (iv) a socialist united states of Europe, both post-NATO and post-free-market.[30] The discussions opened up the lines of possibility on offer and simultaneously underscored the unease of the existing UK political elite because the continuation of the post-war certainties of NATOism/free worldism looked decidedly unlikely.

The early British debates about the EEC were limited in scope. In hindsight it was a feeble debate: the intellectual baggage of the past weighed heavily and there was little discussion of the unfolding dynamics of change. However, the terms of debate were changed irrevocably by the events of 1989/91. The EU response took the form of a renewed drive towards union; occasioning further confusions amongst the British political classes.

Later British debates in the 1990s

As the long interregnum[31] unfolded, debates within the UK public sphere changed: familiar debates were joined by new preoccupations – sovereignty, culture and the broader issue of the UK's geo-political location (Europe versus America). All these debates were sharpened by the 2003 Iraq crisis.

The issue of 'sovereignty' came to be treated as a talisman – keeping it was good, losing it was bad. However, when the nature of the 'it' was examined, debate, such as it was, dissolved away. The member countries of the EU have been 'pooling sovereignty' for many years, and the British government, in the main, had acquiesced in these moves. If these debates are examined, then against the sceptical public argument that sovereignty is somehow given, unitary and inalienable it can be pointed out that the term simply designates patterns of power and authority – which institution, constituted how, responsible to whom, does what. In this sense sovereignty is dispersed through the machineries of the EU, its constituent nation-states and their sub-regional elements.[32]

The debate about culture reproduced the pattern. A distinctive, separate, indivisible and unique Britain/Britishness was contrasted with the mainland – 'Brussels' – in order to argue that change meant loss. Change might mean loss to chronic nostalgics, but to most change simply means change. There has been extensive domestic change – new class configurations, patterns of consumption and social mores. One might also note that from the *Empire Windrush* onwards there has been a significant inflow of migrants. The UK now has large settled minority ethnic communities.[33] In the case of Britain/Britishness, the current pressure for change is structural; as global/regional patterns reconfigure, a response is required – denial will not suffice, and contrasting a unitary Britain with a monolithic mainland and claiming the former was under threat from the later is simply foolish.

Over the 1990s a new version of an old preoccupation was voiced. What had been taken for granted was now explicitly addressed, namely the business of geo-political location, the way in which the UK, or more precisely the British elite, fitted into the global scheme of things. The notion of Atlanticism was reaffirmed. The priority of the link with the USA became one more talismanic issue – any alteration was presented as loss (and any consolidation or extension as gain). Against this, one commentator[34] has suggested that the only people interested in the 'special relationship' are spies, submariners and prime ministers (concerned respectively with signals intelligence, missile targeting and photo-opportunities).

The relationship of the British to the project of the EU was thrown into sharp relief in the episode of the Iraq war. The failures of the political elite were manifest: the decision to disregard EU partners and rally to the support of Washington; the production of letters of support for Washington (organizing the periphery against the core); and the use of the 'French card' in order to buttress domestic support on the eve of war.[35] The upshot of all this was a division within Europe, the old/new Europe divide.[36]

The debate has had one central strand – anxiety. The project of the EU would be problematical for any government (it is a hugely demanding project), but for the British political elite it is a threat. The Eurosceptics have played on these anxieties. It is true that complex change is unsettling, but change is not new, nor can its contemporary demands be evaded. As Norman Davies[37] notes, the Isles have been home to diverse political-cultural projects. The project of Britain/Britishness is relatively recent, and has shifted and changed over its short history. It is clearly changing now. A new pattern is taking shape and one aspect might be noted: the idea of Britain could be giving way to the idea of England.[38]

Structural reconfigurations and implied changes[39]

The character of future political-cultural projects can only be sketched – the business of social science is interpretive and critical – however some speculations can be made.[40] As structures change, so will ways of understanding. The British political classes operate now within an internationalized global system, with significant regionalization and the dominating presence of the EU. There will be both necessary responses (changes in thought/action which are simply given as soon as changes in the world are recognised[41]) and contingent responses (changes in thought/action which flow from measured reflection upon change[42]). It is also clear that these agent responses will occur in the whole range of social locations (individual, group and institutional/organizational).

As Europe integrates there will be 'winners' and 'losers' as different class fractions read and respond differently, thereby securing different outcomes. There will be a series of ways of reading and reacting to the developing European space: the poor and unskilled, routine white collar, educated professional, business elites, financial elites and those with money.[43] There will be key locations for agent group responses, and the groups so located will thereafter disseminate ideas to the wider community (a process fraught with the possibility of multiple conflicts as a new local UK variant of a new European contested compromise is established). One might speculatively identify the following: the realm of the technical bureaucratic linkages within formal state machineries (the interactions of bureaucrats); the sphere of formal political institutions (the exchanges of politicians); the commercial activities of firms integrated at a European (and maybe thereafter global) level; the linkages of the modern media/academic/art scenes (the dialogues of intellectuals); and the vast potential at the popular level with migrant employment, tourism and the realm of trans-European sport/media/pop culture (the spheres of ordinary life and popular culture).

As an ordered political collectivity, as a nation, then there is likely to be a more or less rapid reworking of social practices, institutional structures and legitimating ideologies. As individuals we would experience local-level reordering as the UK system modulates into Europe. It is likely that these matters will be awkward.

Implied changes in structures of power

The British state is strongly centralized. Any reconstruction of the institutional form of the state runs the risk of opening up demands for its wholesale democratization, and the London-based political elite has shown little sign of being prepared to entertain such issues. Any movement towards federalism represents a considerable threat, not merely to institutional arrangements, but also to patterns of official self-understanding. It has been argued that changes were implied by the federal system that had been latent within the development of the EU over its entire career: institutionally embodied power and authority would be relocated both upwards to the new trans-national EU system and downwards to the newly empowered regions of the UK.[44]

If federalism was one anxiety, sovereignty was another. It was noted[45] that members of both political parties had expressed fears and that abandoning myths of parliamentary sovereignty would open up unpalatable questions about the basic character of the political system of the UK.[46] The re-ordering of familiar structures of power would also entail a related process of ideological reorientation from Atlanticism to Europeanism. David Marquand argued that a new European settlement would inevitably borrow from the successful model of Germany – a further source of anxiety for the British.[47]

These matters can be further considered in regard to the 'devolution' introduced by the New Labour government. Tom Nairn characterized it as defensive,[48] yet there is now a parliament in Scotland and an assembly in Wales. Norman Davies[49] reports that opinion polls suggest most of the UK population expects independence for Scotland at some stage. In both cases the newly empowered regional governments have operated differently from Westminster – mild changes, but changes none the less. It is also clear that London has resisted these changes, and as regards English devolution the saga of the choice of first Mayor of London was widely taken as symptomatic.

Implied changes in structures of production

The consistent ploy of the British government has been to play down the importance of the EU, to deny that anything very dramatic was happening. The whole business of the reorientation of structures of production, and the way in which agent groups have read these structures, is problematic for the elite. In recent years commentators have spoken of three clear variants of industrial capitalism: Anglo-American competitive liberal markets; Rhineland social-market corporatism; and the East Asian developmental model. The idea of the diversity of modes of industrial capitalism has offered one framework within which the trajectory of the UK economy could be considered.

The appropriateness of a shift away from the American free-market model to which the British ruling groups have looked over the post-war period, notwithstanding the efforts of the subordinate classes to entrench Keynesian-ordered welfare corporatism, has been underscored by the present changes within the global economic structures. The detail of these reorientations is complex and in brief they comprise: (i) movement on a global level with the development of an extensively internationalized sphere of interaction (economic, primarily – most obviously in finance but also in services involving intellectual property and in trans-national company manufacturing); (ii) the emergence of three major regional groupings in North America, East Asia and Europe (which impacts on the UK as patterns of trade will shift in the direction of one of the blocks); and (iii) movement within the European grouping where there is an emergent European economic space (where this increasingly embraces the UK economic space). The upshot of all these processes of structural change is to shift the UK economic space away from the Atlantic sphere towards the European sphere. If this notes patterns of activity, then there is a wider economic/cultural aspect, namely the shift from Anglo-Saxon competitive markets to European social markets. The way in which the business of economic activity, and thereafter the proper role of the state, is construed within the Anglo-Saxon tradition is quite different to the way these matters are set up and dealt with in the social/Christian democratic traditions of the mainland. The nub of the matter is that whereas the Anglo-Saxons take the market to be a reality *sui generis* and amenable only to rule-maintaining protective intervention by the state, the mainlanders lodge economies within societies and construe the community-serving role of the state in more active terms.

Implied changes in cultural structures

The increasing interpenetration of the nations subsumed within the European Union will have far-reaching effects on the ways in which citizens of the various countries read their relationship to the polities of which they are members. Political-cultural identity is carried in routine social processes: patterns of economic, social, political and cultural activity. Political-cultural identity is layered, made up of various elements. The idea of nation has been one layer – arguably a master status – but Europe is now emerging as one further layer. It is a matter of partial reorientation from the national unit to some sort of European entity. One might anticipate local-level reorientations in a number of areas: formal political structures, institutions and law, how the polity works;[50] and patterns of social identity, how one understands one's location.

The UK has a great tradition, an official realm of ideas/institutions. Thereafter, the UK acquiescent little tradition is the prime vehicle of public identity for the population. This inward- and backward-looking tradition, with its celebration of the small-scale remembered past (home) might be expected to open up to European contexts – we can point to economic links, travel, communication and media, and a similarity of consumer style and patterns of life between the UK and the mainland. In this instance Englishness could not be assimilated into the empire ideology of Britain/Britishness simply because that idea would be in eclipse. A new direct statement of Englishness would have to be made. It would be a notion of Englishness within the context of Europe.

Informal and formal identity claims are lodged in developing time, we inhabit continually reworked tradition. Ordinarily one might expect the reworking of tradition to be accomplished within the frame of an expectation of continuity, of things going on pretty much as they always have, but the shift into Europe entails a disjunction, a new phase in the political-cultural development of the UK. It might thus be taken to be unusually fraught issue. It is certainly true that there is a wealth of available commentary on the notions of Britain/England, mostly the former, but it is clear that the later issue is now on many research agendas.[51]

First debates – first agendas (narrow, implausible?)

The UK had a distinct position within the post-Second World War system: economically, the loss of the empire block was catastrophic and long-term relative economic decline has slowly shifted towards absolute decline in certain key sectors (in the 1980s this decline has been masked by receipts from oil). Politically, the role of America's number one ally in Europe together with the moral status of being one of the victors in the war against fascism has allowed the British ruling class to avoid difficult decisions about the place of the UK in the wider world.[52] This posture of official nostalgia has also served to keep the population quiet. The extent of the investment of the UK polity in the post-Second World War settlement should not be underestimated, however that arrangement is now overthrown by events, and questions in respect of the fundamental nature of the polity can be addressed.

In the public sphere debate has been impoverished. There has been much posturing and successive governments have issued (in effect) threats: either the move into Europe was inevitable, so accept it, or resistance would have severe negative consequences, so accept it. The Eurosceptics have rightly pointed out that threats do not amount to debate, much less persuasion. An argued case for Europe has not been made nor, it might be said, has a reasoned negative case. None the less, the expectation must be that the UK will be absorbed into Europe. Mainland economic, social and political styles will come to the fore.

Marquand[53] has argued that there was a particular legacy to the English revolution. The political class, which apparently responded successfully to the industrial revolution, have proved unable to deal with the subsequent changes in their global-system circumstances. In the late nineteenth century the political class effectively locked itself into an intellectual, political and institutional posture that ensured relative economic decline. In contrast, the Germans, Americans, Japanese and northwest Europeans variously contrived to establish the political, institutional and cultural bases of developmental states. Perry Anderson[54] speaks of 'European social democracy'. It is a distinctive political-cultural tradition: (i) 'citizenship' – there is routine and acknowledged membership of society in contrast to the UK situation of elite/mass; (ii) 'public sphere' – there is a realm of citizen participation in political decision-making in contrast to the UK where the elite rule and the mass are passive/demobilized; and (iii) 'civil society' – there is a realm of private and family activity-in-community in contrast to UK tendency to privatized consumption.[55] The threat to the British political elite is severe. It is entirely unsurprising therefore, that the matter of Europe has come to assume a central place in UK politics.

Notes

1. H. Young 1999, *This Blessed Plot: Britain and Europe from Churchill to Blair*, London, Macmillan.
2. This section is derived from P. W. Preston 1994, *Europe, Democracy and the Dissolution of Britain*, Aldershot, Dartmouth, pp. 130–7. I use the term 'national' after the argument of P. Wright 1985, *On Living in an Old Country*, London, Verso.
3. D. Marquand 2003, *New Statesman* March.
4. The technical term 'dexis' points to the use of the 'we' that includes the audience in without making such a move explicit, for example when elite members address larger audiences using the 'we'.
5. The post-Second World War period has seen the UK political elite accommodating to American priority. A 'US liberalism' (enshrined in their constitution) has resonated with domestic UK liberalism, the legacy of Hobbes/Locke, theorists of the successful agrarian/mercantile bourgeoisie. As argued here, the UK elite has a series of strands of thinking: liberal, conservative, oligarchic. It is a shifting package (and not enshrined in a constitution).
6. This can be called 'British exceptionalism' (as with the US version). It finds expression in routine habits of the state, such as the claim to authority and the requirement of obedience and gratitude, seen most recently with Prime Minister

Blair – 'trust me, I'm Tony'. From any politician this is simply ludicrous, as it contrasts with an appeal to evidence, insight and argument in the public sphere, that is, democracy. Yet in the UK the claim resonates with structural features of the polity and established habits of elite address to the subaltern population. Thus Blair was listened to (for a while), rather than being laughed out of court.

7 B. Laffan, R. O'Donnell and M. Smith 2000, *Europe's Experimental Union: Rethinking Integration*, London, Routledge.

8 See W. Wallace 1990, *The Transformation of Western Europe*, London, Pinter; J. Lodge (ed.) 1989, *The European Community: The Challenge of the Future*, London, Pinter; J. Richardson (ed.) 2001, *European Union: Power and Policy Making*, London, Routledge.

9 S. Bulmer 1992, 'Britain and the European Community' in S. George (ed.) *Britain and the European Community: The Politics of Semi-Detachment*, Oxford University Press.

10 N. Ashford 1992, 'The Political Parties' in George (ed.).

11 Bulmer 1992.

12 The original 1957 group included France, Germany, Belgium, the Netherlands, Luxembourg and Italy. A second group in 1973 included the UK, Ireland and Denmark. Greece joined in 1981, Spain and Portugal in 1985, and the last group to join were Austria, Finland and Sweden in 1995.

13 See D. Urwin (1997, *A Political History of Western Europe Since 1945*, London, Longman), who discusses the UK government's EFTA strategy of organizing the periphery against the core. Its failure was acknowledged in Macmillan's application to join the EEC

14 Bulmer 1992, p. 29.

15 G. Edwards 1992, 'Central Government' in George (ed.).

16 Edwards 1992, p. 67.

17 Bulmer 1992.

18 Urwin 1997.

19 See Young 1999 – Heath was pro-Europe.

20 S. George 1992, 'The Policy of British Governments within the EC' in George (ed.).

21 George 1992.

22 George 1992.

23 It was all presented in business-economics terms. P. Cecchini 1988, *The European Challenge 1992: The Benefits of a Single Market*, London, Wildwood.

24 Roger Backhouse has noted that the economic theorists' recourse to mathematics in the post-war period was in part to escape from the demands of politicians (personal communication) – clearly politicians have learned the trick. See also Roger Backhouse 2002, *The Penguin History of Economics*, Harmondsworth, Penguin.

25 H. Overbeek 1990, *Global Capitalism and National Decline*, London, Allen and Unwin.

26 P. Hirst and G. Thompson 1999 (2nd edn), *Globalization in Question*, Cambridge, Polity.

27 Marquand 2003.

28 J. Pinder 1989, 'The Single Market: A Step Towards a European Union' in Lodge (ed.); J. Pinder 1991, *The European Community: The Building of a Union*, Oxford University Press.

29 J. Palmer 1988, *Europe Without America*, Oxford University Press.

30 A. G. Frank 1983, *The European Challenge*, Nottingham, Spokesman Books. The debates implied a series of wider issues, some of which Frank addressed. In regard to West/West conflicts he made a series of points: (i) that there were conflicts over economic policy, a matter of assertion of interests, and those of the USA were not the same as Europe's; (ii) that there were similar squabbles over international trade; (iii) that there were similar problems over the relation of North to South as the latter was an area of resources vital to both; (iv) that there were conflicts over trade with the Eastern block where the USA took a much harder line than mainland Europeans; and (v) that there had been conflict over the role of NATO, with the Europeans seeing no Soviet threat and the USA keen to keep its very useful and profitable enemy in being. With these problems in mind Frank pointed to a series of economic policy strategies that had been tried: (i) beggar thy neighbour competition, the basis of the squabbles noted earlier; (ii) monetarism, as tried in the USA and UK; (iii) supply-side stimulus, as tried in particular in the USA with Reaganomics; (iv) supply-side re-industrialization policies, inspired by Japan, recommended by US democrats and tried by France in the early 1980s; (v) Keynesian pump-priming as tried by Chancellor Schmidt and President Mitterand, and as proposed by the British Labour Party alternative economic strategy (AES); and (vi) the global Keynesianism of the Brandt Report. It was a confused pattern. However, Frank saw one optimistic line of development, a post-block 'Fortress Europe' freed from the USA.

31 T. Judt 2002, 'The Past is Another Country: Myth and Memory in Post-war Europe', in J. W. Muller (ed.), *Memory and Power in Post War Europe*, Cambridge University Press.

32 A comparative note might be made here: on the mainland power and authority are (often) dispersed through the political structures of the nation-states, while in Britain the political structure revolves around the Whitehall/Westminster village.

33 Y. Alibai-Brown 2001, *Who Do We Think We Are: Imagining the New Britain*, Harmondsworth, Penguin, but see also Zadie Smith or Hanif Kureshi. Are these now not minorities, but more mainstream contemporary English?

34 R. Braithwaite 2003, *Prospect* March/April.

35 On this see, Marquand 2003.

36 In more narrowly focused analyses, others characterized Prime Minister Blair as a weak leader, blocked at home in respect of Europe (in particular the euro currency issue), choosing to become engaged in adventures overseas (the 'war against terrorism' serving as a displaced object for otherwise blocked labourist emotivism – passion, commitment, etc.). The crisis revealed the seemingly instinctive preference of the British political elite for the USA in preference (or in this case, against) Europe.

37 N. Davies 2000, *The Isles: A History*, London, Papermac.

38 N. Davies 2001, 'Britain and Australia: Holding Together or Falling Apart?' Public lecture at the *New South Wales Centenary Federation Committee Symposium , The Holding Together Program.*

39 The section is derived from Preston 1994, pp. 144–50.

40 As with the 'scenarios' of chapter 6.

41 Recall Peter Winch on (say) the invention of germ theory and consequent medical practice. P. Winch 1958 (2nd edn 1990), *The Idea of a Social Science and its Relation to Philosophy*, London, Routledge.

42 Recall Alasdair MacIntyre on ideology – choices are made and this set of ideas is pursued rather than some other. A. MacIntyre 1962, 'A Mistake about Causality in Social Science', in P. Laslett and W. G. Runciman (eds), *Philosophy, Politics and Society, Series 2*, Oxford, Blackwell.

43 The impacts of complex change might be more obvious to the members of some groups than to others, or to persons considered simply as individuals, for example farmers (with the rule of the CAP) or academics (with the ERASMUS schemes) or business people (with EU regulations).

44 D. Marquand 1988, *The Unprincipled Society*, London, Fontana.

45 W. Wallace 1990, *New Statesman and Society*, 9 November.

46 Wallace 1990.

47 Marquand 1990.

48 T. Nairn 2002, *Pariah*, London, Verso.

49 Davies 2001.

50 One thing the various inhabitants of the UK will have in common is the switch from being subjects of the crown-in-parliament to being citizens of a European Union – a first experience of citizenship.

51 R. Porter 1992, *Myths of the English*, Cambridge, Polity; A. Easthope 1999, *Englishness and National Culture*, London, Routledge; P. Longford 2000, *Englishness Identified: Manners and Character 1650–1850*, Oxford University Press; R. Colls 2002, *Identity of England*, Oxford University Press.

52 See C. Hitchens 1990, *Blood, Class and Nostalgia: Anglo-American Ironies*, London, Vintage.

53 Marquand 1988.

54 P. Anderson 1992, *English Questions*, London, Verso.

55 J. Keane 1988, *Democracy and Civil Society*, London, Verso.

8 The European Union project

The European Union[1] has a particular historical occasion, a distinctive institutional matrix, a spread of relationships with the wider world and a developing collective identity.[2] The present situation – 'what there is' – is not easy to characterize, and nor is the likely path of future development, but the EU might best be read as a contingent accumulation of institutional machineries which together vehicle a nascent polity.[3] An irregular dynamic of change has shaped the historical development experience of the EU: specific contexts have shaped the concerns of actors, who have formulated lines of advance, which have found expression in institutional machineries and various strategies of giving practical effect to these intentions. Thereafter, multiple lines of commentary have been advanced, feeding into further occasions for agent action. The cycle continues;[4] accumulating institutional competencies building around a contested core project, the ideal of 'ever closer union'.

The EU project: context, actors and practice

The middle years of the twentieth century saw a series of wars,[5] with various participants, each having different experiences of chaos, death and destruction. A series of regional impacts can be identified: in East Asia, there was extensive warfare followed by slow reconstruction; in Europe, there was similar destruction and slow recovery; and in sharp contrast there was economic and political advance in the USA, Latin America and Africa. The locations of these regions, and their constituent countries, within the global system were radically altered, and their experiences feed into the post-Second World War period and contemporary memories in quite different ways.[6]

In Europe these experiences have shaped the construction of the EU. The European experience of the twentieth century was one of profound violence.

In the Great War European societies experienced a great shock, not merely the brutal novelty of industrialized warfare,[7] but the moral impact upon their self-understandings as pinnacles of civilized achievement. It brought considerable soul searching,[8] but in the event little was achieved. In the 1920s and 30s Europe was a mix of established empire powers, the recently defeated and various new nations plus national minorities. It was an unstable pattern, further disturbed by the rise of the Soviet Union,[9] the contested Treaty of Versailles,[10] the ineffectual League of Nations[11] and failed revolutions in Germany followed by Weimar and the multiple conflicts of the 1930s.[12] The Second World War can be read as a further episode in what has been called a 'general crisis'.[13]

The extent of the destruction in Europe is difficult to comprehend:[14] the Great War killed some eight million soldiers, and the Second World War killed some forty million people, with the bulk of the casualties being civilians.[15] The war saw extensive forced relocations of peoples: national socialist relocations/murders from 1938/45, and in 1944/45 a further series of dislocations/relocations as the Allied armies secured their victory.[16] The war saw extensive material damage to the infrastructure of European countries:[17] cities, towns, industrial plants, communications networks and cultural treasures. The war in Eastern Europe was particularly savage, with correspondingly high levels of death and destruction.[18] By 1945 the continent was divided and occupied, in the East by the Soviet Union and in the West by the USA. In addition, through the war years, the colonial holdings of the various European countries had been either occupied,[19] or promised help in exchange for post-war independence,[20] or had re-oriented their economies and international links towards the USA.[21] At the end of the war the idea of empire was untenable, despite the ambiguous behaviour of the USA and some ill-considered European attempts at recovery (or reconquest). It is in these circumstances that we find the impulse to the project of the European Union.

European reconstruction began within the framework of American plans for a liberal trading regime. There were two key sets of institutions: the UN organization and the Bretton Woods system, plus the confection of the cold war. The Bretton Woods system was organized around the IBRD, IMF and GATT and created a liberal-market trading area. It was successful, with rapid economic recovery in Western Europe, the USA confirmed as the core of the entire sphere and the USSR excluded. The complementary political apparatus centred on Washington, the UN and the machineries of cold war. The UN provided the formal framework of a liberal regime,[22] nation-states and law. The cold war buttressed these arrangements. The standard Western story presents the cold war as a response to potential Soviet aggression, although Gabriel Kolko[23] points out that the US government did not merely react to Soviet activities but began planning early for the post-war period. At the same time in the USA there was a critical domestic politics: the Republican Party's response to Roosevelt's New Deal was hostile, an aspect of this was electoral red-baiting (which under Roosevelt did not count for much but which did under his successor Truman, who let it all run). The cold war comprised an extensive economic, diplomatic, military and ideological competition that legitimated European division, disciplined allies and helped order the liberal-market West (block-time, block-thinking and block-leaders).[24] So, by 1945, the US had both an economic and political strategy. Russian anxieties with regard to security and the shape of post-war Europe ensured that exchanges with the USA and other Western allies would be difficult. In the event, relations deteriorated rapidly. Europe was divided. The western part was absorbed into an American sphere, the eastern into the Soviet. It is in this political environment that the project of the EU began: at first in the modest form of the European Coal and Steel Community (ECSC), thereafter the European Atomic Energy Community (Euratom) and the EEC.[25]

The political elites of France, Germany, Italy and the Benelux countries were the key actors in the early development of the EU.[26] There was a complex interchange, a process of mutual re-working of elite self-identities and their images of Europe.[27] It is a lengthy tale.[28] In the immediate post-war period the French political elite had no coherent idea of Europe, or how to deal with Germany. A period of domestic confusion and colonial warfare was brought to a close with the accession of President de Gaulle and the 1958 founding of the Fifth Republic. A vision of France as a model of rational enlightenment civilization was constructed. The related model for Europe was one of nation-states. The general stance continued with President Pompidou. A fuller accommodation with Europe was achieved only slowly: the 1980s crisis of President Mitterand's Keynesian-style socialism in one country was one turning point, the end of the cold war another. The elite image of France was expanded to embrace the EU; commentators speak of the Europeanization of French exceptionalism. In contrast with the slow adjustment of the French political elite, the German elite (and people), lead by Chancellor Adenauer, embraced Europe as a route away from a catastrophic national past into a democratic future. There were other views, but these were fairly quickly absorbed into the pro-EU line.[29] The position was adhered to throughout the post-war period, and the consensus was not disturbed by the subsequent 1990 drama of reunification. In the cases of Italy[30] and the Benelux[31] countries similar stories can be told: the European project was quickly embraced – the commitment of the core member is strong. However, in sharp contrast, the British elite preferred the consolations of a continued pretence to an international role. Their resistance to the European project was fixed in place very early.[32]

The post-war history shows how quickly the elites of Europe committed themselves to the project of Europe – they began early and they have persisted. The trajectory of the EU has not been smooth however; it can be tracked and phases identified. The overall sequence of phases is recognisable in the detailed stream of events, the treaties and the development of the institutional machinery of Brussels.[33] The 1950s were successful, the 1960s/70s experienced Euro-sclerosis, the 1970s/80s saw renewed advances and the 1990s/2000s have seen dramatic advances. The EU has a powerful, increasingly integrated, economy, deepening cross-national social networks and the outlines of a nascent polity. It has also accumulated a dense institutional structure – executive, civil service, parliament and court – which deeply interpenetrates the constituent national systems and serves to order the overall political-cultural project.

Decisions/institutions, phase 1: the Treaty of Paris and the Treaty of Rome

In the chaos of the early post-war years, the political elites of mainland Europe addressed the task of centring European politics on cooperation rather than war. The major actors were the political elites of France, Germany, Italy and the Benelux countries. The key figures were Jean Monnet and Robert Schuman. A series of decisions were taken that established crucial institutions: the first move came with the Treaty of Paris (1951) that set up the 1952 ECSC. The rationale was clear: establish supra-national control of the strategic and economically interlinked industries of coal and steel (the keys in the 1950s to modern economies and war-making capacity) and thereby foster peace. The machinery of the ECSC anticipated the subsequent structures of the EEC/EU: an element of law-based supra-national authority with extensive lower level consultations amongst functional experts. Thereafter, there were further moves. The political issue of security was acknowledged when the French government in 1952 proposed a European Defence Community. It was accepted by the other members, but failed in 1954 to gather support in French Assembly, a failure that produced a desire for movement elsewhere. The key actors were once again Monnet and Schuman. A proposal from the Benelux countries for economic integration was adopted and translated into practice by the 1957 Treaty of Rome. The ECSC machinery was revised, providing the keys to the political project and institutional structure of the EU (council, commission, parliament and court). The primacy of EU law and the principle of its direct effect were established in 1963/64 by the European Court of Justice. However, there were problems: President de Gaulle, affirming the priority of the nation-state, resisted supra-national decision-making. The crisis was settled with the 1966 Luxembourg Agreement that provided for the effective reintroduction of unanimity in the council and the Committee of Permanent Representatives (CORPERER), that is, effectively, the national veto. One might say, broadly, that the original scheme combined 'high politics' and 'low politics' in a productive way. The supra-national elements were high politics, but they were undemanding at the outset, the rest was the low politics of technical economics questions. The EU developed in this fashion: elite-level decisions mixed with low-level mundane advance.

The development of the EU has involved commentary – it has been a thoroughly well-discussed process. The early debates revolved around federal/functional positions, and were shaped by the character of twentieth-century European history. In the interwar period there had been discussion of a federal Europe, a utopian political project. In the post-war period this was reintroduced, but there were many practical difficulties, however it did generate what came to be called the Monnet method. The strategy focused on technical/functional issues overseen by a law-based supra-national high authority, and invited participants to proceed slowly, anticipating spill-over effects, thereby buttressing cooperation. David Mitrany[34] addressed these themes in terms of functionalism, looking not to ideals and programmes but to the technical functions in the modern world. He argued that experts should take charge and order a series of networks located above the state, and attacked the idea of European integration as statist. The neo-functionalist approach can be seen in the work of Ernst Haas[35] who argued for an incremental approach to European integration, the Monnet method. Influenced by 1950s US functionalist social science, Haas focused on processes of integration: economics is the key whose influences spill over into other functional spheres, the social and the political. The theory – which has been called the 'authorized version' of European integration – is elaborate. The concern for processes of integration unpacks in a number of ways. It shows concern for building institutions, turning ideas into law. Institutions are seen as agents of integration, vehicles to transfer elite loyalty to the European level (interaction of belief systems). The role of social actors and experts is acknowledged (national elites and experts). It assumes that Europe would move to the centre of actors' activities, whether they are for or against, and the role of spill-over effects is crucial. The approach was used to vehicle the integrationist project.

However, neo-functionalism was subject to criticism, both intellectual and practical. European social science of the 1950s and 60s was strongly influenced by American work; the ideas of functionalism and the preference for positivism were prominent. These intellectual predilections had been subject to extensive criticism, but practical problems (social conflict rather than harmony in the 1960s for example) led to the collapse of functionalist social science, and neo-functionalism looked exposed. The coincidental arrival of de Gaulle in Europe, strongly nationalist, reaffirmed the nation-state and functionalist ideas lost favour. The confusions in commentary mirrored the wider political situation as overall the EU project lost impetus in the 1960s.

Decisions/institutions, phase 2: Euro-sclerosis and expansion, to the mid-1970s

The early 1970s saw significant changes in the circumstances of the EU project. In the West there was a falling away in US pre-eminence (a consequence of the Vietnam war) and a general economic slowdown (a consequence of the failure of the Bretton Woods system of fixed-currency exchange rates). In the countries of the EU there was a new generation of leaders: Charles de Gaulle left office after the events of Paris '68 and was succeeded by George Pompidou, thereafter Giscard D'Estaing; in Germany Konrad Adenauer was succeeded by Ludwig Erhard, Kurt-Georg Kiesinger then Willi Brandt; while Edward Heath became Prime Minister of the UK. The European Commission gained a new leader in Roy Jenkins.

The UK – having been rebuffed earlier by de Gaulle – renewed its application for membership to the EU, and with Ireland and Denmark joined in 1973 (the government of Norway agreed membership but lost the subsequent referendum). A series of proposals for institutional reform were made: the idea of monetary union (EMU, a first response to new situation of floating exchange rates); and the idea of European political cooperation (EPC, a response to pressures of Vietnam, Ost-politik coupled with internal dynamics). However, there was little immediate success. Later, President Giscard d'Estaing successfully proposed the establishment of the European Council, and direct elections to the European Parliament were also agreed. Once again, progress was slow, but there was progress. At the time commentators spoke of 'Euro-sclerosis' but the period later came to be more favourably judged (with low-level work and interesting commentary, in retrospect). The period is typical of the EU: complex reforms, crab-like advances.

The 1970s Euro-sclerosis attracted significant comment. There were many putative symptoms: de Gaulle's actions inhibited change, the failure of Bretton Woods introduced instability and inflation and whilst there were new EU members little seemed to be happening. A new theoretical stance was produced, in which Stanley Hoffman[36] reaffirmed the role of the state. The intergovernmentalist approach rejected neo-functionalist supra-nationalism (with its accumulative advance, spill over and so on), and argued that the development of the EU was not merely the result of the activities of states but was also dependent upon their continuing engagement. In related work it was suggested that integration was one instance of the interdependence of states.[37] This line produced a more complex analysis of the role of the state and the nature of internal/external interdependencies. Finally, a liberal intergovernmentalist position[38] was presented that read the EU as an intergovernmental bargain based on the needs of powerful economic interests.[39] At this point the familiar debate between supra-nationalists and intergovernmentalists took off. It ran for some considerable time, though its value is unclear.

Decisions/institutions, phase 3: the SEA and the TEU

The failure of the Bretton Woods system signalled changes within the global economy. The American economy experienced problems with trade, industrial restructuring and debt. In Europe commentators spoke of Euro-sclerosis. As petro-dollars were recycled, problems of debt grew in Latin America and Africa. However, other parts of the world were more successful. In East Asia there was rapid economic advance; in Europe there were new initiatives. Jacques Delors argued for the completion of the single market, a long-term community goal. Albert Spinetti presented a draft federal constitution in the European parliament. Greece in 1981, Spain and Portugal in 1986 became members of the EU. These initiatives signalled renewed movement within the union, a new phase in its historical development. The 1987 Single European Act committed members to the completion of the single market, reforms to community institutions and made the European Council part of the core machinery of the EU. Then, in 1989/91, as the domestic reform programme unfolded, the wider international situation changed.

The post-Second World War system was suddenly fluid. A host of questions were raised for Europe: German reunification was possible; the role of the USA was unclear; and there were dramatic changes in East and Central Europe. A group of key actors, including Delors, Mitterand, Kohl, Bush and Gorbachev, confronted the task of remaking the post-Second World War settlement in Europe. As the former Eastern block dissolved, a series of prospective EU members emerged, NATO was expanded[40] and the task of the reform and expansion of the union within an increasingly globalized[41] world was addressed. An Intergovernmental Conference (IGC) opened the way to European monetary union and European political union, and the decisions were enshrined in the 1992 Maastricht Treaty of European Union (TEU), which established the three institutional pillars of the union,[42] the euro currency and the European Central Bank (ECB). A series of new members joined in 1995: Austria, Finland and Sweden, and a further IGC was held to consider further institutional reforms, leading to the 1999 Amsterdam Treaty and the 2000 Nice Treaty. In 1997 accession discussions began with countries in Central and Eastern Europe and the 2002 Copenhagen European Council agreed the accession of Poland, the Czech Republic, Hungary, Slovenia, Malta and the Baltic states. In 2003 a group chaired by Giscard d'Estaing published a contentious first draft of a constitution.

The period was dramatic and attracted much commentary. So, first, there was remarkable practical activity. The old project of 'ever closer union' surged forwards with the 1985 SEA and the treaties of Maastricht 1992, Amsterdam 1997 and Nice 2000. Second, there was new theoretical commentary:[43] (i) the established debate between neo-functionalism and intergovernmentalism looked dated and restricted; (ii) the ideas of complex policy making and multi-level governance appeared; (iii) there came a new concern for the role of institutions; (iv) the idea of policy network analysis was developed;[44] and (v) the idea of 'deep regionalism' was presented.[45] Overall, material drawn from international relations and cast in positivist terms fell away and there was a slow drift to seeing the EU as a polity with commentators asking how it all worked, rather than what it was. As positivist work faded, and political science and international relations borrowed from sociology, there was much talk of 'social contructivism'.[46] This generated a new issue: how is Europe being made?

Decisions/institutions, phase 4: the implications of 2003

The EU at Copenhagen 2002 concluded accession discussions with a range of former Soviet block countries and Mediterranean states, and further expansion of Europe was signalled. The process promised to increase to twenty-six the number of member countries. The prospect of administrative confusion had been recognized, as had the requirement for the EU to attend to its long-standing 'democratic deficit', and both issues were pursued in the context of the proposed EU constitution. These matters would have been awkward at the best of times, however the Iraq crisis provoked divisions. The European political elites split, concerned with not merely the war, but also the character of the Bush regime, the relationship of the EU/USA and the nature of the European project. The mainland core countries of France and Germany opposed the war and asserted the distinctive interests of Europe: law, multi-lateralism and consultation. The British state-regime rallied to the support of the USA, as did the political elites of Spain, Italy, Poland and other Eastern European states. The division was cast in terms of old/new Europe,[47] but the crucial division, once again, was between the British, endeavouring to mobilize the periphery against the core, and their mainland partners.[48] The promise of Copenhagen was tarnished but crucial issues in respect of the character and future direction of the EU were out in the open.

The machinery of the EU[49]

It is possible to identify a series of phases in the development of the EU. In each, there are patterns of structural change and lines of agent response. It is a shifting, contingent process that has been subject to extensive analysis. The materials produced by social theorists, one part of the accumulating weight of commentary, reveal shifting centres of attention.

The overall debate is extensive. There have been numerous context-bound attempts to theorize Europe.[50] As we have seen, in the pre-war days there were proposals for a federal Europe, a utopian political movement (still extant) seeing integration as better than conflict. Mitrany argued for the transfer of power to experts who could oversee the functional necessities of modern life – another reaction to war, a politics of no-politics. In the post-war period these debates were revisited and represented in new forms: neo-functionalists focused on the mundane details of integration (the 'authorized version' of European integration; and a resurgent intergovernmentalism insisted on a return to high politics and the role of the state. Ben Rosamund suggests that the 1990s saw renewed theoretical lines of advance, and following the SEA and TEU attention shifted to ideas of complex policy-making and multi-level governance. The EU has come to be treated as an odd sort of polity.

The analytical focus on the Brussels machinery has produced its own debates. In the mid-2000s the dominant models for analyzing policy-making are the ideas of policy communities and policy networks,[51] both species of an actor-oriented approach to social scientific analysis. The overall strategy looks like this: (i) adopt an actor-oriented approach; (ii) try to track just what is happening (never mind trying to fit reality to the expectations of big positivistic theories); and (iii) then take note that the EU has lots of actors, lots of networks and lots of institutions (where we find decision points). An elaborate analytical vocabulary is available. First, the idea of actor-oriented analysis: pay attention to the patterns of relationships between actors, the ways in which they understand, the ways in which they act and interact, and the ways in which policy statements/actions are the outcome of long, complicated, untidy processes of interaction. Second, the idea of a policy community: an image of a settled group of policy-makers (roughly, in a modern European state, a group of a few thousand) with discussions of policy largely internal to this group who thus set general political agendas and guide the taking of formal decisions. Third, the idea of policy network (domestic/EU): the idea of a policy communities began with a national focus, however, the EU involves a series of national elites whose work intermingles – the idea of networks can grasp this wider intermingling of people. The networks are fluid and rather open, many people join in and patterns of interaction are not as fixed and familiar as they are within former national elites. Fourth, the related idea of an epistemic community: communities of experts, those with technical trainings, play a distinctive role within the wider policy network. They are players, not providers of neutral expertise. Finally, the notions of advocacy coalitions and policy streams: once we have the image of policy networks, the issue becomes one of inaugurating and running lines of policy: the idea of advocacy coalitions addresses the first issue and the idea of policy streams the second. It is not a neat process; it is loose, fluid and messy. This gives us a general picture: epistemic communities (experts) plus advocacy coalitions (interested parties) combine to feed ideas into the Brussels policy community/network (debates amongst relevant Brussels groups), who may in time formulate ideas which can be fed into the next stage, the processing and exchanges in the formal institutions of the EU coupled with parallel intergovernmental bargaining, and from this a decision emerges, shaping policy, law and institutions. It all gets translated into practice, changes the general situation, and loops back to the thinking of experts and advocacy groups – the cycle continues.

The concern for process can operate at more general levels. Laffan *et al.*[52] discuss both questions – what is it and how does it work – and offer an answer to the former via a discussion of the latter. They identify three phases in the development of the EU – (i) Bretton Woods; (ii) neo-liberalism; and (iii) post-1989 – and argue that events have outrun theory. In order to advance matters, the authors make four key assumptions: (i) the contingent nature of integration; (ii) the variability of integration, it is not unidirectional; (iii) the routine intermeshing of internal/external dynamics; and (iv) the historical trajectory is important. And to this they add two key ideas: (i) 'territory, identity and function' (the nature of states today); and (ii) 'ties and tensions' (the dilemmas of states/markets, polity/policy, EU/global and EU policy/multi-levels of policy). In sum, the approach has three elements: (i) interconnected contingent historical processes; (ii) complex inter-relations of territory, identity and function; and (iii) an appreciation of the tensions associated with the process. The authors argue that individual European states are becoming embedded in wider economic and political systems. The international political economy is changing: the global economy has multiple networks and multiple actors. World politics are changing, at global, regional and national levels. The end of the cold war revealed these changes, and the EU must act within these new systems, it must adapt. The authors suggest that Brussels is weak, but knowledgeable (indeed, it may be that Brussels has leaned lessons others must). The EU is adjusting domestically: governance is evolving, accreting a dense institutional network as national and European levels intermesh; markets are integrating within the EU, where the euro is a significant step, and in the global system. The EU begins to look like a polity. In sum, the EU is responding, effectively and slowly. All this gives us a new model of EU integration and character: (i) it is robust and innovative; (ii) the national state is weakening or relocating (global, regional or sub-national); (iii) there is extensive national/EU inter-embedding; (iv) the EU is not federal, it is not intergovernmental, rather it looks like 'deep regionalism'; and (v) it is changing, it is an ongoing project, thus integration is deepening, flexible and contingent (there is no overall blueprint, what has been constructed could have been constructed otherwise and might in the future be reconstructed).

The EU political-cultural project in overview

The debate amongst theorists and commentators about the nature and dynamic of the EU is ongoing. There are a series of broad lines of analysis: (i) neo-functionalism, generating influential ideas of spill over; (ii) intergovernmentalism, an international relations perspective derived from neo-realism, which stresses the role of state (state and market versions are available[53]); and (iii) governance and policy-making, a social science approach which looks at political processes, rather than formal institutions and law. A series of points about these debates might be made: (i) there is an obvious mix of scholarship and politics (that is, engaged political commentary); (ii) there is a clear mixture of American and European work (an interesting mixture, though the appropriateness of using American resources, which are shaped by culture and tradition just as any other social science, to grasp an emphatically European project, must be in some doubt); and finally (iii) the problem of theorizing an ongoing project - while the EU continues to develop, so too does the body of accumulated theorizing and commentary.

At present, in the mid-2000s, the practical situation could be summarized as follows: (i) the elite commitment is strong (and popular support and acquiescence is also in place), the memory of war, chaos and occupation remains strong and central; (ii) the development trajectory of the project is contingent, as is its current pattern, and the final character and destination of the project is unclear and contested; (iii) the institutional vehicle(s) of this project are (now) elaborate – the whole machinery of the EU – interpenetrating both national and European levels and continuing to develop; (iv) the pattern of development has been driven by internal dynamics but has been shaped by changing global circumstances (the early context for the development of the EU was the political/economic project of 'the West', dominated by the USA and more recently the EU has begun to make clear its own concerns and agendas); and(v) it became clear in 1989/91–2003 that the EU was both up and running and central to Europe's future. Also at present, the state of intellectual debate could be summarized as follows: (i) the EU is not a super-state in the making; (ii) it is not an international organization; (iii) it is a response to demands of the internationalized global system and the internal dynamics of its members; and (iv) in terms of complex change, one could speak of a series of national trajectories contriving a deep regional sphere, thus the EU has many institutional structures and many voices but thus far it has moved towards 'ever closer union'.

The European Union as a global actor

The political coherence of the EU is growing. The countries of the union constitute a powerful economic grouping; the EU machinery is central to this power. The countries have social, cultural and political linkages throughout the global system, and the union's linkages with East Asia, the Third World and the USA are well established. The EU is an established global actor, however it remains the case that it is both militarily and politically weak.

Institutional reform, expansion and the CFSP

The end of the cold war signalled the close of a chapter of European history. A series of dramatic events signalled change: Gorbachev's domestic reform programmes; citizen groups in the Eastern European countries; the sequence of changes in autumn 1989; German reunification in 1990; the Soviet block dissolving in 1991; the end of the block system; the early mis-celebrations of Western commentators; the end of familiar ways of thinking. This last required that matters long evident be acknowledged (in particular the economic and political power of the EU) and implied that many existing institutions would have to be remade. The changes being considered and undertaken have been complex and grasping their logic is difficult. Yet after the initial reactions, debate about the shape of Europe quickly came to revolve around the EU. It was a ready-made political-cultural project. It is firmly established in Brussels and the member countries. The EU members have extensive links throughout the global system.[54] The role of the EU itself is limited, but it has developing relationships with the various component parts of the global system (on matters involving politics, policy and the construction of institutional vehicles of recognition and co-operation, in particular in regard to economics and trade). As the EU moved towards the centre of debate, the members had to think about their political role; this raised awkward issues. As this role was acknowledged, three areas of activity were important: the expansion of various non-EU European organizations; the expansion of the EU; the development of the EU Common Foreign and Security Policy (CFSP). In the mid-2000s the implications of debates are far-reaching and securing agreement on understandings and actions is both highly contested and political.

A series of largely Western cold war organizations have expanded their membership in the years following 1989.[55] Such expansion and reconstruction is bound up in contested visions of the future of Europe and the EU. We can take the example of NATO, a crucial organization as it projects the influence of the USA into the heart of European politics. NATO was founded in 1949 with twelve members (the USA, Canada, the UK, the Benelux countries, Denmark, France, Italy, Iceland and Norway), and its establishment was one move in the development of the cold war divide between the USSR and USA. The European Defence Community (EDC) was proposed in 1952 and failed in 1954. NATO then emerged as the key instrument of US hegemony within the West and the central organization for European defence. A process of expansion followed: Greece and Turkey joined in 1952; the Federal Republic of Germany in 1955; Spain in 1982. The organization became an integral element of block-time. Now, however, its future is subject to debate.

After the upheavals of 1989 new discussions about enlargement began. These were not straightforward matters. A series of stages can be identified: (i) managing change in the USSR 1989–91 – thus no new members, rather talk of association; (ii) deferring change 1992–93 – the North Atlantic Coop Council acknowledged the concerns of Eastern European countries and managed the anxieties of the CIS; and (iii) new movement 1994–97 – there was evident instability in the CIS coupled with applications from former Soviet block countries to the EU, at which point NATO membership became available. Overall, there were multiple political debates: managing change in the USSR; an internal debate within NATO (about what to do); aspirations from Eastern European countries to return to Europe and be located in a plausible security system (that is, not simply as members of a broad trans-European organization); and there were anxieties about patterns of post-1991 change in Russia. The upshot was that NATO accepted new members, but that there was debate about its role. It was suggested that as NATO had expanded its old role had been left behind, and that the organization was at the end of its useful life ('NATO is dead'). Later, an out-of-area role was proposed and some operations undertaken. Later still, discussions around the 2003 proposed EU constitution saw the French, Germans and British (plus others) agree the establishment of an EU proto-general staff.[56]

As we have seen, the early 1970s was a pessimistic period, with commentary on Europe was dominated by talk of Euro-sclerosis. In the 1980s however came a sequence of enlargements.[57] The individual countries had different motives for membership: economics was crucial, as was returning to Europe, shifting into the democratic mainstream and joining a powerful secure grouping. They also had different expectations of membership: economic advantage, economic advantage and security, or the whole political-cultural package of 'ever closer union'.[58] In addition, the established EU member countries had different motives for such enlargement: for the German political elite, expansion towards the East was a final task in rectifying the consequences of the Second World War; for the French elite, expansion threatened an unfortunate dilution of the project of an independent Europe;[59] and for the British, the expansion seems to have been read as providing new peripheral allies in the fight against the centre.

The diplomatic/military aspect of the developing political role and identity of the EU has been problematical. The TEU established the CFSP pillar,[60] but there were different views on this. The German elite was pro-federalist and so pro-CFSP, but there were problems with historical memory, their constitution and capacity. The French elite was pro-nation-state and so uneasy about the CFSP, but had memories of national greatness[61] and was in favour of the EU having a greater international voice. The British elite was pro-nation-state and so opposed to the CFSP, but had memories of empire, represented in recent years via its pro-USA/NATO stance. After the Yugoslavian wars however, they granted that some EU military capacity was necessary. Other countries had other opinions, some pro-USA, some neutral. In the event, the Iraq crisis showed that the CFSP was little more than declaratory posturing and the member country political elites went their own ways. However a quartet of countries made another attempt at a European defence identity (Germany, France, Belgium and Luxembourg) and the 2003 constitutional talks also acknowledged the security sphere. Overall, as currently there is unsustainable unclarity, the issue will be revisited.

The global system – links with East Asia

The European involvement with East Asia is long established and has had a deep impact upon both areas. The definition of East Asia is self-conscious, pointing roughly to the area of Chinese cultural predominance. The importance of the existing civilizations has been underscored by A. G. Frank[62] who argues that prior to the arrival of the Europeans in the 1600s East Asia was the centre of the global economy, rich and trading across a wide area. The Europeans then entered these circuits of East Asian trade and developed on the basis of this involvement. While such involvement was at first minor – just one more group of traders – it grew over the years until the Europeans were significant players in local economies and politics; the fullest expression of these interests came in the 1800s and 1900s as the Europeans and Americans established formal colonies. In 1941 the European countries still controlled a series of empires. These were fatally disturbed by the events of the Pacific War. Thereafter there was rapid decolonisation and cold war division within East Asia. The USA became the dominant extra-regional power: military, political and economic. The European countries in large measure simply withdrew.[63]

The post-war division of East Asia along cold war lines saw an autarchic socialism in the People's Republic of China and export-oriented development in Japan and the Asian Tigers. The countries of Southeast Asia grew more slowly. The reform programme inaugurated by Deng Xioaping in 1978 allowed China to construct its own version of the East Asian model of development, and the region is now rich (albeit unevenly). The interest of European countries in the region has slowly revived. Individual European countries have maintained links that date back to colonial days: language, trade and flows of migration; the EU has had commercial treaties with China for many years and has agreements with the Association of Southeast Asian Nations (ASEAN). The end of the cold war saw a new concern for the links between Asia and Europe. The countries of East Asia have taken note of the rise of Europe and there have been anxieties about EU protectionism, cast in terms of 'open/closed regionalism' (in other words, the economic linkages have regained something of their former importance). There have also been debates about tri-regionalism, the emerging relationships between Europe, American and Asia. The anxieties of the East Asian nations, the importance of trade and the discussions about post-cold war global architecture generated the proposal to establish a dialogue between Europe and Asia: the Asia-Europe Meeting (ASEM).[64]

The ASEM process provides a formal vehicle for the countries of the EU to meet bi-annually with the countries of East Asia. The initiative came from the government of Singapore, and a series of meetings has since been held. It is difficult to specify precise results from these meetings; they are perhaps best regarded as an institutional acknowledgement of a deep mutual interest, rather than anything further. In the meantime the various member countries and the EU itself continue to deepen existing trade linkages, in particular with Japan, China and the East Asian Tiger economies.[65]

The global system – links with the Third World

The European colonial empires held territories in Africa, Latin America, the Middle East, South Asia, Southeast Asia and China. The Second World War undermined the system, however extensive links were retained albeit sharply revised in the new political situation of decolonisation and the establishment of new nationstates. The early expression of the reworked linkages came in the 1975 Lome Convention, whereby the EU gave preferential access and aid to former colonies. Over time this came to be seen as less than successful. New ideas about development emerged and a new agreement was made that shifts the focus to encouraging trade; the 2000 Cotonou Agreement.

The engagement of Europe with the Third World is historically long established. In the years following the Renaissance, as Europe prospered it opened trade links with various parts of the global economy; these links in time turned into formal and semi-formal empires. The involvement varied but the overall story is one of the slow remaking of extant economies and societies in line with the needs of industrial capitalism, a mixture of destruction, exploitation and development.[66] The Second World War destroyed the political possibility of sustaining formal empires: the disruptions of war, broad changes in political opinion and the rise of indigenous nationalist movements meant that the European holdings could not be re-established. However, the economic interests in markets, resources and trade that had lead to the establishment of colonies remained.

The relationships of metropolitan capitalism and various peripheral capitalisms were reworked over the post-war period. The key ideas were independence and development, which acknowledged changed political circumstances and continuing linkages. The early discussions of development took place in the context of both decolonization and cold war. The tasks of rebalancing the relationships and setting in place independence and development, were made more awkward by the block competition of the cold war era.[67] The European countries maintained bi-lateral relationships with former colonies and adopted common positions in the context of the EU. The member countries undertook to extend development assistance to the countries of the Third World, in particular former colonial territories. The early strategies were shaped by contemporary ideas about the process of development: preferential access and industrialization were stressed, and found expression in the Rome treaties, Youande declarations and the Lome Convention. In more recent years as ideas about development have changed and the Lome strategy has come to be seen to have failed, a new approach has been adopted. The Cotonou Agreement concentrates on trade.

The EU links with the Third World have been called a policy patchwork.[68] Development policies (trade, aid and foreign direct investment plus other social/political programmes) are Pillar One, whilst the CFSP is Pillar Two. The formulation of development policy has been difficult as most of it is foreign-policy motivated. The events of 1989 complicated things with a shift of attention from the south to the east, and WTO rules added further complications. The EU contributes over and above the work of the members. Lome and Cotonou are more public, egalitarian and committed than the actions of member governments. The EU needs a development policy as a matter of self-interest, as the Union is not hermetically sealed and has extensive linkages already.[69]

The global system – the USA

In 1945 at the end of the Second World War the USA was the dominant world power. It had the greatest concentration of industrial power, the greatest concentration of financial power, the largest, most technically advanced armed forces, and offered a cultural model to many people in Europe and those countries looking either to recover from war or to escape from colonial rule. It was the start of the popular image of the USA as a rich consumer democracy.

The wartime political elite of the USA had been influenced by its own experiences of interwar economic depression and the disruptions of the Second World War. The dominant elite fraction argued that a liberal trading sphere ordered by the USA would best guarantee prosperity and thus avoid conflict. The institutions of the global system with which we are familiar were constructed at this time and with these ideals in mind: the United Nations, the World Bank, the GATT (later WTO) and the IMF. Overall, the idea was to regulate the global economy and regulate global politics. In the US sphere there was stability and great prosperity. All this lasted until 1971 when the costs of running the Vietnam War meant that the USA had to modify the Bretton Woods system. The international system of fixed exchange rates was ended, and the global financial system entered a period of floating exchange rates. In 1973 the first 'oil shock' injected inflation into the system, and by the 1980s the idealism in the West in respect of regulated economies and regulated politics began to be replaced by ideas favorable to free market competition – the New Right package and its ideas of deregulation, liberalization and globalization. However, what had happened in this period, as was made clear by the end of the cold war, was that the USA had declined relative to other nations. It was no longer the preeminent power. There were now three regions in the global economy: each had about one-third of the global GDP, each had a powerful financial sector, each had a powerful knowledge base and each had its own sense of itself as a cultural region.[70] In the years since, the implications of 1989/91 have slowly been uncovered and discussed. The rebalancing of EU/US relations has been a subtle and evolving process, manifest in a series of institutional locations, hence debates about trade, the environment, the international criminal court and defence. It has also been manifest in the changing nature of the relations of individual EU members with the USA. Matters were brought to a head following the 9/11 attacks and the subsequent wars in Afghanistan and Iraq.

In the USA the first few months of George Bush's presidency were marked amongst other things by public indiference to the Middle East.[71] Then came 9/11. There were alternative views of 9/11: it was evidence of the clash of civilizations;[72] it was radical Islamist terrorism;[73] it was 'blowback';[74] and it was a local-level response to US involvement in the Middle East. These debates are ongoing amongst commentators and scholars. The US political/military elite quickly formulated a policy: 'the war against terrorism'. It was a narrow response – a mix of revenge, demonstration and opportunism – and as events unfolded, the problems and multiple implications became evident. The first expression of this response was the attack on Afghanistan, which viewed charitably was understandable but unfortunate.[76] The second was the war with Iraq, widely condemned as illegal, unjustified and reckless.[77] The regime was contained,[78] but American neo-conservatives saw an opportunity for unilateral preventative war (and beyond that the project of democratizing the Middle East). In Europe, while the TEU had made new structures, affirmed new aspirations (through the CFSP) and opened the way to new members, the Iraq war produced a divided response, with individual national elites responding differently. One set were pro-American – the UK's Blair, Spain's Aznar and Italy's Berlusconi offered support for a war without UN authority. Another set opposed the Bush government – France, Germany, Austria and Belgium defended ideas of multilateralism and the role of the UN. The divide was characterized as 'new/old Europe'. The confused response is indicative of the deeper post-cold war problem of rebalancing US and EU interests.[79]

If we ask how these events play within the dynamics of the EU, the answer is that significant short-term damage has been caused though, in the medium term, it may underline the importance of common European projects. The immediate impact of the Iraq crisis is in high politics – the new/old Europe divide – but in low politics, discussions about EU business have continued. In popular politics the anti-war movement was Europe-wide. Germany and France were the motor of Europe – now maybe restarted. The British were semi-detatched – now maybe disregarded. More importantly, the events have revealed the longer term issues: where does power in the EU lie; what is the direction of the EU; what is the location of the EU within the global system? The comfortable political-cultural territory of the West is dissolving. The US has its own interests, as does the EU.

The EU as a global actor

In the mid-2000s, the EU now has a significant presence as a global actor. The economic weight and historically accumulated linkages of the countries of the EU with regions around the globe secure this position. In brief, the EU is already a major world economic power. It has also been extensively involved in development aid and various initiatives to build international law and community. However, it is also clear that the collective weight of the EU is not well represented within the global system. The EU remains politically weak and this is likely to persist until a common foreign and security policy is established.

The European Union as a polity – a developing identity

The development of post-war Europe was not merely a matter of institutional apparatus and political movements but was also in significant measure a cultural process. In external matters, Western Europe was subsumed within the American sphere, the interwar economic and political failures in Europe were forgotten, as were the successes in the Soviet Union as the cold war ethos reaffirmed the vigour of liberalism (in markets and polities). The domestic construction of a new idea of Europe involved creating an agreed version of the recently ended war, a complex mix of active and selective[80] 'remembering and forgetting'.[81] Tony Judt tracks the moves. The war was very destructive – the chaos had multiple aspects (physical, social and political). An official version of the war quickly took shape: it was all the fault of Germany in general and the national socialists in particular; there was a modest period of purges after which Western Europe pursued free markets and Eastern Europe pursued socialism. The details of the immediate aftermath of the war – the fluidity, the choices made about how to remember/forget and go forwards – have been forgotten. Over time the official Western version came to be questioned. The claims have slowly been revisited, in particular in recent years around the idea of memory. The official Wastern version came up for grabs after the world of official state socialism collapsed and a new episode of forgetting (that is, the period of state socialism) looked likely. After 1989/91 there was an interregnum as the old pair of block-time truths were gone.[82] Since the interregnum ended in 2003, there is little sign of a new agreed or official truth. It might be that none is needed (after Jurgen Habermas, it is better that scholars debate these matters and present truths in the public sphere), but it is clear that 'Euro-chat' is not enough.[83] None the less, it can be suggested that the history of Europe will have to be revisited as the EU project advances, as any new perspective is likely to coalesce around this project. All this has a wide spread of implications for the ways in which individuals and groups understand themselves: ideas of nation, ideas of Europe and received expectations of the future are in process of reformulation.

The notion of identity can be grasped in various ways. The trio of ideas introduced earlier – locale, network and memory – offer a way of grasping the rich detail of ordinary life. Richard Hoggart[84] was concerned to analyze the forms-of-life and understanding of ordinary people in their ordinary lives. It is in these routines that we make our lives; it is here that we have our homes, friends, neighbours, jobs, hobbies and so on. It sounds rather obvious, but until Hoggart (and others) pointed this out, the whole sphere of life and experience tended to be ignored by social scientists in favour of other agendas (policy advice to government, elite history, abstract economic theorizing, metropolitan political commentary and so on). The approach allows us to think about the fine detail of experience and identity, about the ways in which it routinely depends upon other people, those with whom we variously interact. It lets us see that some aspects of our identity rest on dense interactions and carry an intimate weight of meaning. As we move away from our home base, social interactions become less demanding, they carry less meaning. As we move away from familiar private spheres into the public sphere, other agents contribute more strongly and we experience directly the distribution of power within society. The model of locale, network and identity thus presents identity as layered. The local domestic sphere is rich in meaning. It is a sphere where we have control. As we move away from the private sphere and into the public sphere meaning thins, and our own power becomes less.

It is in the public sphere that we meet official ideas: official truths, ideologies and the extensive routines that sustain nationalism in everyday life.[85] The standard story of nations offered by nationalists (and probably the view that 'runs through our heads'), is that national identity is a given of human life. One might say for nationalists, a national identity is as obvious and natural as a gender identity. However, the social scientific view of all this is that nations (like gender identities) are social constructs, elaborate sets of ideas, carried in key institutions and learned and relearned in ordinary social routines. A new schedule of questions is generated: when were nations invented, why and with what consequences? Barrington Moore[86] discusses ideas of change and shows how the nation-states with which we are familiar developed over the long period of the shift from the world of feudalism into the modern world. Moore uses ideas of historical trajectories, stressing the contingency of patterns of development. As these trajectories unfold, patterns of life and patterns of ideas unfold. One might look to the EU in similar terms: there have been a series of 'European identities' and a new form of European identity might emerge as a long-term historical construction/achievement.

Europe as discourse, past/future

The idea of discourse points to the link between knowledge and power. As social life is disciplined, so too is the production and dissemination of knowledge. As we learn, we are disciplined, that is, drawn into a way of thinking and behaving. All formal knowledge claims are embedded in practice and institutions. In the modern world the state is the major agent of knowledge and power. Foucault offers examples: (i) crime (definitions, law, machinery of law – police, courts and goals – and popular obedience); (ii) madness (clinics etc.); and (iii) sexuality (disciplining the body).[87] Using these ideas, Gerard Delanty[88] argues that the EU is one particular, contingent and situated version of Europe: particular as it makes certain claims, to democracy, to free markets, to human rights and so on (it could be another set); contingent in that it has no essence, it is the out-turn of complex processes and we can sketch its social scientific history (its economics, society and politics); and situated as it has emerged in a particular historical and social scientific context and it is shaped by that context. More broadly Delanty points to five historical discourses of Europe: (i) the medieval discourse of Christendom; (ii) the enlightenment discourse of civilization; (iii) the late nineteenth- and early twentieth-century discourse of culture; (iv) the cold war discourse of the West; and (v) the current contested discussions centred on the EU.

After 1989/91, the project of the EU came to the fore in European political life. The nature of the EU is unclear: there is an ideal of 'ever closer union' but no common understanding of that goal; the identity of the Union, its component institutional parts (including nation-states) and its peoples is not settled; the character and direction of the Union are subject to intense debate.[89] It is likely that an EU identity will develop but it will only emerge over time. One might speculate that 'European-ness' is likely to be one more layer of people's identities.[90]

Notes

1 The treatment here is firmly non-specialist. There is now a large literature on the EU, and here the project is reviewed on the basis of the work of contemporary scholars. It is important for non-specialists – in particular in the UK – to have a rough idea of the vigour of the project.

2 In the course of the domestic debates about the wisdom or otherwise of the Prime Minister's decision to actively support the Bush regime's Iraq adventure, David Marquand summarized the ethos of the European Union: 'The European Union is not primarily about economics, or even politics in the ordinary sense. It embodies a vision of transnational governance that springs from experience of conquest, reconquest, slaughter, genocide, dictatorship and torture during the first half of the twentieth century. The Union's founders … had lived through that experience … The Union they built has nothing in common with previous attempts to unite Europe by force. It is based on law, and it is quintessentially multi-lateralist. It is also based on the tacit premise that its members give a higher priority to their common enterprise than to their relations with any non-European power.' He adds 'In siding with the unilateralist United States against the multilateralists of old Europe, Blair shows that he does not understand – and, still worse,

that he does not care – what the European Union is for.' D. Marquand 2003, *New Statesman* March.

3 B. Laffan, R. O'Donnell and M. Smith 2000, *Europe's Experimental Union: Rethinking Integration*, London, Routledge. The authors offer the idea of 'deep regionalism'.

4 Diagramatically: context > actors > decisions/institutions > practice (> commentary) > new context.

5 Thus: 1914–18 Great War; 1928–47 Chinese Civil War; 1931/37–45 Sino-Japanese War; 1938/39–45 the Second World War; the 1941–45 Pacific War; and, thereafter, the 1947–89/91 'cold war'.

6 See, for example, John Dower 1999, *Embracing Defeat: Japan in the Aftermath of World War II*, London, Allen Lane; C. Thorne 1986, *The Far Eastern War: States and Societies 1941–45*, London, Counterpoint; M. Mazower 1998, *Dark Continent: Europe's Twentieth Century*, New York, Alfred Knopf; A. Mayer 1988, *Why Did the Heavens Not Darken: The 'Final Solution' in History*, New York, Pantheon; R. Prebisch 1950, *The Economic Development of Latin America and its Principal Problems*, New York, United Nations; and in the USA a mythologized version of the Second World War is available, looking at the material and moral aspects of the US victory. See W. I. Hitchcock 2003, *Prospect* April.

7 N. Davies 1997, *Europe: A History*, London, Pimlico. Davies gives figures for deaths and both military and civilian casualties.

8 For example, the rise of International Relations. See E. H. Carr (1939) 2001, *The Twenty Year Crisis*, London, Palgrave. Alternatively see F. R. Leavis and the idea of the redemptive power of culture through English literature – see C. Jenks 1983, *Culture*, London, Routledge.

9 Made central by E. Hobsbawm 1994, *The Age of Extremes: The Short Twentieth Century 1914–1991*, London, Michael Joseph.

10 On Versailles, for an alternate perspective, see M. Macmillan 2003, *Peacemakers: Six Months that Changed the World*, London, John Murray.

11 See Carr (1939) 2001. On Carr see C. Jones 1998, *E. H. Carr and International Relations: A Duty to Lie*, Cambridge University Press; J. Haslam 1999, *The Vices of Integrity: E H Carr 1892–1982*, London, Verso.

12 See Hobsbawm 1994 on this period of crisis. On the Third Reich, see W. L. Shirer 1960, *The Rise and Fall of the Third Reich*, London, Secker and Warburg.

13 Mayer 1988.

14 See Davies 1997; Mazower 1998.

15 Thanks to the invention of 'air power' and 'total war' See S. Lindquist 2002, *A History of Bombing*, London, Granta; W. G. Sebald 2004, *On the Natural History of Destruction*, Harmondsworth, Penguin.

16 This included the expulsion of a large number of ethnic Germans from Eastern Europe. This particular aspect of the wider catastrophe has recently formed the subject of Gunter Grass' 2003, *Crabwalk*, London, Faber and Faber, see also G. Grass 2000, *My Century*, London, Faber and Faber.

17 Estimates vary, but GDP was reduced to the levels of 1900.

18 A point noted by Mayer 1988.

19 Thus: French Indo China, Hong Kong, Malaya and Burma, the Dutch East Indies and a scatter of islands in the Pacific otherwise held by Europeans and Americans.

20 India.

21 Thus: Latin America. The situation in Africa has one interesting witness in the form of the Martha Quest novels of Doris Lessing.

22 Analysed in IR terms as 'regime theory' whereby sets of transnational rules govern the interaction of states and market players.

23 G. Kolko 1968, *The Politics of War*, New York, Vintage. The people who did the planning were shaped by the experience of depression and war and were convinced that the answers to the problems were to be found in economic prosperity carried by economic liberalism, in particular a global free-trade regime.

24 Schematically there are three positions on cold war: (i) the multiple conflict was Stalin's fault (J. L. Gaddis 1997, *We Know Now: Rethinking Cold War History*, Oxford University Press); (ii) it was the fault of Republican red-baiters plus Truman's weakness (Kolko 1968); and (iii) it was a mechanism of control (R. Aron 1973, *The Imperial Republic: The US and the World 1945–1973*, London, Weidenfeld). I take aspects of the last two – the first seems foolish.

25 For a chronology see R. MacAlister 1997, *From EC to EU: An Historical and Political Survey*, London, Routledge.

26 The key actors were the national political elites, though the detail here is beyond my scope. The EU is the outcome of these complex national/European politics, and the present pattern could be described in numerous ways as cores/peripheries or as a series of blocks. The original core of six, the Scandanavians, the Meditteranean, and Central European countries, left the British as an outlier. This opens up the issue of geographical metaphors for the dynamics of the EU (also not pursued here).

27 M. Marcussen *et al.* 2001 'Constructing Europe? The Evolution of Nation-State Identities' in T. Christiansen *et al.* (eds), *The Social Construction of Europe*, London, Sage. Marcussen accesses their political elites via party elites; it is thus a restricted notion of elite, but does discuss the public politics.

28 Marcussen *et al.* 2001, pp. 104–14.

29 H. James 1994, *A German Identity: 1770 to the Present Day*, London, Phoenix, chapter 9. See also V. R. Berghahn 1988, *Modern Germany: Society, Economy and Politics in the Twentieth Century*, Cambridge University Press, pp. 197–216.

30 P. Ginsborg 1990, *A History of Contemporary Italy*, Harmondsworth, Penguin, p. 160. See also W. Wessels, A. Maurer and J. Mittag (eds) 2003, *Fifteen into One: The European Union and its Member States*, Manchester University Press, chapter 11. They argue that the EU has offered the Italian elite and masses a model against which to measure, order and orient their own society.

31 Wessels *et al.* 2003.

32 H. Young 1999, *This Blessed Plot: Britain and Europe from Churchill to Blair*, London, Macmillan.

33 MacAlister 1997.

34 D. Mitrany 1966, *A Working Peace System*, Chiacgo, Quandrangle Books; D. Mitrany 1975, *The Functional Theory of Politics*, London, Martin Robertson.

35 E. Haas 1958, *The Uniting of Europe*, Stanford University Press; E. Haas 1964, *Beyond the Nationstate: Functionalism and International Organisations*, Stanford University Press.
36 Stanley Hoffman 1966, 'Obstinate of Obsolete: The State in Western Europe', *Deadalus* 95.
37 R. Keohane and J. Nye 1977, *Power and Interdependence, World Politics in Transition*, Boston, Little, Brown.
38 Thus, the state and the market can be set up as opposed, but in liberal line there are two ways in which contracting individuals can find security, either in political or market structures, two ways of controlling and ordering power (finally a fundamentally asocial characteristic of individuals). See P. W. Preston 1994, *State, Market and Polity in the Analysis of Complex Change*, Aldershot, Avebury.
39 A. Moravcsik 1998, *The Choice for Europe*, London, UCL Press.
40 S. Croft, J. Redmond, G. Wyn Rees and M. Webber 1999, *The Enlargement of Europe*, Manchester University Press.
41 B. Rosamund 2001, 'Discourses of Globalization and European Identities' in Christiansen *et al.* (eds).
42 EEC, CFSP and justice (the last two are intergovernmental).
43 B. Rosamund 2000, *Theories of European Integration*, London, Palgrave.
44 Rosamund 2000.
45 Laffan *et al.* 2000.
46 Christiansen *et al.* (eds) 2001.
47 The Spanish and Italian prime ministers were both weak figures who supported Bush against the wishes of their populations. The Eastern European countries have different historical trajectories and thus their elites and populations have different perspectives, however the episode of the letters revealed the naivety of the elites in regard to EU politics.
48 Marquand 2003.
49 The EU organization:

European Council

Council of Ministers   Commission   European Parliament

Court of Justice   Court of Auditors

Economic Social Committee   Committee of Regions

50 Rosamund 2000.
51 J. Richardson 2000, 'Policy making in the EU: Interests, Ideas and Garbage Cans of Primeval Soup', in J. Ricardson (ed.) *European Union: Power and Policy Making*, London, Routledge.
52 Laffan *et al.* 2000.
53 The root is liberalism, which can unpack in two ways – a focus on the state (secures order) or the market (secures order). Thus the IR debates focus on neo-realism versus neo-liberalism. See C. Hay 2002, *Political Analysis*, London, Palgrave, chaper 1; also B. M. A. Crawford 2000, *Idealism and Realism in International Relations*, London, Routledge. Crawford argues that the pursuit of a scientific discipline (i.e.

scientistic and separate) distorted the whole business and advocates going back to E. H. Carr as a species of critical theorist.

54 C. Pienning 1997, *Global Europe: The European Union in World Affairs*, Boulder, Lynne Reinner.

55 Croft *et al.* 1999.

56 A European defence identity requires that Europeans discuss and decide upon common problems and threats; in turn this implies an institutional vehicle for these conversations. On these matters I am indebted to conversations with Terry Terriff.

57 UK, Denmark and Ireland in 1973; Greece in 1981; Spain and Portugal in 1986; EFTANS in 1995 (Austria, Finland, Sweden, Norway and Switzerland (last two chose not join)); the CEECS in 2004 onwards (Poland, Hungary, Czech Republic, Slovenia, Estonia, and later Slovakia, Bulgaria, Romania, Latvia and Lithuania) and the Mediterranean countries (Malta, Cyprus and Turkey).

58 The procedure starts with a letter of application to the Council of Ministers. The Commission comments and if all is well detailed negotiations begin. The average length is around five to six years. The applicants must accept the body of Union law and negotiate transitional arrangements.

59 See Marcussen *et al.* 2001.

60 The TEU set up the CFSP as one of its three pillars. It was a mix of intergovernmental elements and some supra-national ones – thus, heads of state (in the form of the European Council) decided and passed these decisions to Council of Ministers to execute. There were two elements to the decisions, common positions and joint actions. The latter included humanitarian aid, observer missions to elections, sanctions against Zimbabwe elite; the former included declarations. One key problem is that what is a common position and what is a joint action is vague, for example over finance (who pays). The scope of all this is limited and the overall set up is unclear – the judgement is that it will all have to be revisited.

61 T. Nairn (2002, *Pariah*, London, Verso) comments that French 'grandeur' is analogous to British post-imperial posturing.

62 A. G. Frank 1998, *Re-Orient: Global Economy in the Asian Age*, University of California Press.

63 P. W. Preston 1998, *Pacific Asia in the Global System*, Oxford, Blackwell.

64 P. W. Preston and J. Gilson (eds) 2001, *The European Union and East Asia*, Cheltenham, Edward Elgar; see also J. Gilson 2000, *Japan and the European Union*, London, Macmillan.

65 Pienning 1997.

66 See P. Worsley 1984, *The Three Worlds: Culture and World Development*, London, Weidenfeld.

67 See F. Halliday 1989, *Cold War, Third World*, London, Radius.

68 M. Holland 2002, *The European Union and the Third World*, London, Palgrave.

69 Holland 2002.

70 The USA has an extensive global military deployment, though it is not clear why – the USA does not look much like an empire. However Chalmers Johnson (2000, *Blowback: The Costs and Consequences of American Empire*, Boston, Little, Brown) argues that the US military itself has quasi-empire aspirations. See also M. Mann 2003, *Incoherent Empire*, London, Verso.

71 As the 2003 presidential election campaign began there were claims from former Bush cabinet officers that Iraq had been a target from day one.

72 In this paper he argues that after the cold war there will be no more ideological conflicts rather there will be conflicts rooted in deep cultural differences (and Huntington spoke of the deep cultural differences between the West and everyone else). The solution is for the West to work together to resist inevitable conflicts.

73 These groups do not represent Islam, they do not represent any social movements in Middle Eastern or Islamic countries, they represent only themselves – they are irrational fanatics. The solution is to destroy the terrorists (plus maybe reform and develop the countries they come from so that fewer radicals will emerge in the future).

74 Johnson 2002 argues that the US political/military elite has interfered in a series of local conflicts over the years and one result is local hostility to the USA (the expression 'blowback' comes from the CIA, for example, Osama bin Laden was supported, funded and armed by the USA when they were backing a war in Afghanistan against the forces of the Soviet Union.) The solution is for the USA to stop getting involved in local conflicts.

75 The US has stationed troops in Saudi Arabia and supported local regimes – in Saudi, Egypt and Israel. The solution is to readjust American foreign policy in respect of the Middle East.

76 P. W. Preston 2002, '9/11: Making Enemies; Some Uncomfortable Lessons for Europe'. Paper presented to the *European Union in International Affairs Conference*, Australian National University, 3/4 July.

77 In the UK the view ran across the political spectrum – published criticisms are available from (for example) David Marquand, Corelli Barnett and Tariq Ali.

78 J. J. Mearsheimer and S. Walt 2003, 'An Unnecessary War', *Prospect* March.

79 As the Iraq crisis unfolded through 2004 the political distance between the EU and the USA grew: a series of textx examined the hubris of the US regime (Mann 2003; E. Todd 3003, *After the Empire: The Breakdown of the American Empire*, Columbia University Press; C. Johnson 2004, *The Sorrows of Empire: Militarism, Secrecy and the End of the Republic*, London, Verso); others the misjudgements of Prime Minister Blair (N. Cohen 2003, *Pretty Straight Guys*, London, Faber and Faber; J. Kampfner 2003, *Blairs Wars*, London, The Free Press; see also R. Cook 2003, *The Point of Departure*, London, The Free Press).

80 On memory – this crops up later, but we note personal, group and collective memory. We can identify an hierarchy as we shift from the small-scale, detailed and personal to the general schematic and finally to the 'official national past'. Memory, the sets of ideas running through our heads, is a crucial aspect of identity. If we ask how to investigate/reflect on memory, we can ask whose memories (power and agenda setting), we can ask about remembering and forgetting (what gets put in and what gets left out), and we can then revisit (unpacking the metaphors/logics of the story told) and check the claims against the scholarly record. Memory is not a passive receptacle, not simple reportage. Memory is active, a part of the way the social/political game is played. In evaluating memory, a naïve way would be to check the facts, but this will not do. Thereafter, there are two contemporary strategies: first affirm a postmodernist irony – there is no single truth, merely perspectives, so relax, view the matter amiably and with irony (Richard Rorty's suggestion – see also Billig's comment on Rorty: M. Billig 1995, *Banal Nationalism*, London, Sage); or second affirm the modernist project, after Jugen Habermas, and so reject fidelity to the facts and reject utility for the nation, and instead look for the regulative ideal lodged in language (the aspiration to scholarly truth – cognitive and moral) and thereafter the issues are pursued dialogically (it is open-ended).

81 T. Judt 2002, 'The Past is Another Country: Myth and Memory in Post-war Europe' in J. W. Muller (ed.), *Memory and Power in Post War Europe*, Cambridge University Press.

82 Judt 2002.

83 But see T. Diez 2001, 'Speaking Europe: The Politics of Integration Discourse', in Christiansen *et al.* (eds) – talking about Europe is a part of making it.

84 Richard Hoggart 1958, *The Uses of Literacy*, Harmondsworth, Penguin.

85 Billig 1995.

86 B. Moore 1966, *The Social Origins of Dictatorship and Democracy: Lord and Peasant in the Making of the Modern World*, London, Allen Lane.

87 For a brief overview of Foucault's work, see J. G. Merquior 1985, *Foucault*, London, Fontana.

88 G. Delanty 1995, *Inventing Europe: Idea, Identity and Reality*, London, Macmillan.

89 Marcussen *et al.* 2001: (i) 'liberal nationalist' (compatible with a Europe of the nation-states); (ii) 'Europe as a community of values'; (iii) 'Europe as a "third force"' – an alternative to block ideas; (iv) 'a modern Europe as part of the Western community' – liberal-democracy and social market, such as Germany and recently France; and (v) 'a Christian Europe' – a catholic stream of thought, increasingly integrated into iv (p. 105).

90 There are two familiar rhetorical positions: the nationalists fear that the EU will overwhelm existing nations, whilst the federalists hope precisely that it will. Both are likely to be disappointed. A minimum EU of law and principle can be expected coupled to great diversity. One thinks of a Habermasian 'constitutional patriotism', see J. Habermas 1992, 'Citizenship and National Identity: Some Reflections on the Future of Europe', *Praxis International* 12.

9 Storytelling I: discourses of England

The business of reading and reacting to enfolding structural change is a routine part of the life of political communities. It involves a mixture of recognition and choice.[1] In the former if one comes to understand that the world has changed, then all subsequent actions automatically change. In the latter one makes choices – as circumstances change new demands/possibilities emerge, and one route to the future is preferred over another. In terms of the recognition of post-war structural change, and the consequent rise in the importance of Europe, the British elite (and the wider population) have been slow, seemingly always one step behind events.[2] In terms of choices, the relationship of the UK elite and the European project has not been a happy one. The project is a threat to the elite, though it is an opportunity for the masses. One might expect any chosen advance to be slow and indirect, however one might also expect the recognition of given change to proceed whether or not the powerful approve. It is here that we find new forms-of-life and new sets of understandings.

It is now a commonplace amongst scholars that the sometime discrete nation-states of Europe are becoming economically integrated, densely institutionally intermingled and culturally intertwined. It is not a straightforward process. It is not inevitable. It is in this context that the question of the nature and direction (identity) of the ordered political community becomes an issue. In periods of change people become uneasy – familiar domestic settlements are in question; the criteria for judging issues become hazy (part recognition, part choice). The direction is not clear and the future is open to debate.[3] The current changes call into question the familiar post-Second World War settlement, the liberal-democratic national welfare state, and the deeper practice and ideas of Britain/Britishness. The political-cultural project of Britain is no longer tenable. It is neither plausible (it does not grasp extant circumstances), nor useful (it offers no plausible route to the future). The nature and direction of the UK political community is in question, yet the business of reworking received ideas and constituting new identities are not domestic matters for the UK population to resolve in isolation. They will be resolved in the company of wider debates about the nature of Europe. The elaborate exercise in forgetting and remembering that attended the post-war settlement in Europe is also being revisited and agreed truths restated.[4]

So, how might we investigate the relocation of England? We can anticipate change. We can anticipate 'England-in-Europe'. We can speculate about its form. It will centre on practice, changing forms-of-life[5] carrying novel patterns of understanding. It is here that new ideas of England will emerge. But new understandings will begin with available ideas (what else are there?) and in order to move forwards in imagination, we first have to look backwards to review the sets of ideas we take for granted. If we set aside the idea of Britain, we can identify a spread of 'discourses of England', and we can review these resources. Here the materials of the arts/humanities will be introduced, offering a new set of resources; thereafter, we will go on to recall available ideas of England. And with all these resources in plain view, we can turn to the issue of relocating England in Europe.

Culture, identity and journeys

A concern with the ideas carried in routine social practice, the stories we tell ourselves about who we are,[6] is not new.[7] It is the territory of culture.[8] The development of the idea of culture can be tracked [9] and the concept can be seen to be used in a variety of ways.[10] In the nineteenth century a spread of familiar concerns could be identified, for example the nature of modernity, its costs and possible alternatives.[11] The interest in culture mirrored the wider intellectual concerns – the analysis of the changes underway in the shift to the modern world. Such debates ran into the twentieth century, when the social sciences were disciplinized and a series of discrete enquiries emerge. Something similar happened in regard to the sphere of culture – towards the end of the nineteenth century travellers'-tales and the experience of empire found expression as anthropology. English literature became a scholarly concern after the Great War,[12] while cultural studies developed after the Second World War. At this point we had not merely the tales we told about ourselves but a diverse arena of systematic reflection upon the business of telling stories. This material was rich, opening up new approaches, disciplinary resources and substantive issues and concerns.[14]

The sphere of the arts and humanities is large,[15] but it provides material useful to present purposes: thus in recent years, against familiar ideas of identity as fixed, clear and discrete (that is, somehow 'unitary'), it has been proposed that identity is fluid, opaque and multiple.[16] These debates have a series of overlapping strands: (i) 'decentring the self' – that is, rejecting the familiar 'Western individualist' self in favour of a self comprised of multiple relationships with others (a 'self' is thus an amalgam of all those selves presented in the course of everyday life); (ii) 'hybridity' – in the context of colonial holdings the social exchanges of members of different cultures alter both parties – the colonizer and colonized are shaped by the experience, essentialist ideas of self are rejected, selves are amalgams of cultural strands and a common human condition is illustrated; and (iii) 'contingency'- the networks of social relationships which we inhabit are not fixed and permanent, indeed, the reverse – we inhabit a delicate tissue of relationships, bequeathed to us by history, sustained by structural patterns that enfold the communities we inhabit – there is no linear personal or collective progress. At which point it might be objected that if identity is fluid, opaque and multiple, if it is contingent, then what of the common experience of the 'solidity' of our sense of self (stressed in the earlier chapter on identity), and the experience of 'particularity',[18] the world taken for granted? And it is true that most people, most of the time, find identity wholly unproblematical. It is how we live, it is how most of us experience our lives. However, this is only one way of thinking about things. The material of arts and humanities (also the social sciences) offers a way of unpicking this certainty, of revealing how self and identity are social constructs. In routines it does not matter, but in processes of change it does, when what was taken to be secure becomes insecure.[19]

Drawing on the resources of the arts and humanities, taking a clue as to how to proceed, one way of grasping the nature of identity[20] is through the metaphor of journeys. This points to the experience of those who have shifted from one social location to another, those who have shifted from one identity to another – either they have relocated or the world has changed around them[21] – those whose direct experience reveals the contingency of identity. The notion of journeys links identity and learning. In a period of complex change we have to interpret the world from the inside, with what tools we have, and it is likely to be unsettling.[22] The notion of journeys lets us consider the experience of being 'unsettled'. In respect of 'England' we can point to a series of typical journeys: domestic (journeys through a familiar locale); international (journeys along networks to/from England); and reflective (journeys through the available stock of images of England).[23] All the material is detached, one way or another: migrants, émigrés, expatriates and sceptical natives offer the views of knowledgeable outsiders.

Domestic journeys: industry, class and culture

The experience of movement offers a way of making comparisons, and thus of learning the extent and subtlety of a political culture. A journey can be on the part of the observer – actual travel – or it can be relative movement – where the observer records change. The shift to the modern world in the UK offered observers a direct experience of rapid, pervasive, complex change: the rise of industry, the development of new class structures and the claims of democracy all fed into the production of novel cultural reflections.

Journeys through a newly industrial land

Raymond Williams[24] details a series of theorists who offered readings/responses to the rise of industrial capitalism. These nineteenth-century literary figures stressed culture in opposition to the rise of industrial society. A series of thinkers looked at the new industrial world and considered unfolding and prospective political changes. Burke and Cobbett were influential: the former offering a conservative critique of the modern world where the problems were not just industry, but democracy; the latter looking at inequality, recording the impacts of the modern world on the poor, in particular the rural poor, and celebrating an undamaged earlier epoch and the efforts of the poor to organize. Thomas Carlyle also reacted against industrialism and stressed the role of the intellectual in defending culture (understood as acquired skills and sensibilities), which was taken to be under threat by the advance of industry. An organic society had to be defended against laissez-faire society. In a less radical guise, Mathew Arnold stressed learning as a defence against the degradations of industrial mass society, offering sympathy from the middle classes but no action, rather withdrawal.

Others were more active. The shift to the modern world generated new industrial settlements, new patterns of class and middle-class interest in the situation of the urban poor/working classes. A wealth of ethnographic material was produced: it was coupled with calls for social reform and experiments in utopian social reforming. All these materials have a long pedigree and reach back to Friedrich Engels writing on the condition of the English working classes in the 1840s. Such work continued in diverse guises in the twentieth century. The classic modern political text dealing with class in England is George Orwell's *The Road to Wigan Pier*.[25] In respect of the poor, further work has been done by Peter Townsend,[26] and a notable contribution came from the historian E. P. Thompson[27] who rescued the poor, the marginal figures, from 'the condescension of history' and opened up the history of the English working class.

There are other theoretical and practical reactions to the rapid development of industry, thus urban form and social reform were linked. There were attempts to grasp the logic of the industrial world and move forwards, often via an idealized rural past. Rural life was read as clean and wholesome, with rural craft skills read as authentic in contrast to machine-tending in the new factories. The former line can be taken to begin with philanthropic industrialists – Titus Salt (Saltaire), the Lever brothers (Port Sunlight), and the Cadbury family (Bourneville Village) – and finds full influential expression in the early urban planning theories of Ebenezer Howard, the founder of the 'garden city movement', who built Welwyn Garden City and Letchworth. The architectural theme was the denial of the distinction between rural and urban: the provision of gardens, public open space and clean industries would encourage healthy living (both physical and spiritual). The second line, the celebration of life built around rural craft skills, found expression in the arts and crafts movement associated with William Morris, which influenced the early cottages of the architect Edwin Lutyens and Frank Lloyd Wright. Morris linked art and politics, seeking authenticity in both spheres. The concern for urban form, art and social reform continued, running into the 1930s with the Greater London Council, for example, and into the post-1945 period with the new town building.[28]

Patrick Wright[29] displayed the strength of the approach in his analysis of Britain/England through the ethnography of a small area of London (Dalston Lane). The material offers a record of changing social practices and the residues lodged within the urban structure of both these practices and the self-conscious, small-scale interventions that had been made (building, preservation, changing urban uses and so on). Wright traced the ways in which broad patterns of political economic, social institutional and cultural change found expression at the local level; the ways in which changes can run through an area and the ways in which local agents endeavour to read and ride with these changes. The processes of structural change are unending – change is continuous, so too the necessity of responding. However, the whole business is radically untidy, structural changes do not smoothly occasion agent responses, the rhythm is much more broken. At the local level these processes of change and local response are laid down one on top of another as changes run and responses are initiated, run for a while and then die away. The upshot of all this is that reading the urban environment is a matter of diagnosis rather than description, a matter of uncovering layers of action and meaning. In other words the urban environment – people and places – is meaning-drenched and these meanings can only be recovered in a diagnostic or ethnographic fashion.

Journeys through the changing class system

As Williams shows, the theorists of the nineteenth century were very aware of class – the new industrial world was linked to ideas of democracy and evidenced the growth of a large working class. In the twenty-first century UK society continues to be drenched in consciousness of class. A vast literature exists, but two key themes might be mentioned: change and loss.

As post-war prosperity and changing social mores encouraged intergenerational class mobility in the 1950s and 60s the issue found wide expression. The literature of class journeys is a distinctive genre. It is a literature of changing social/status position, either general, along with prosperity, or particular as one person shifts classes (usually from working to middle class, as downward social mobility seems less remarked[30]). These issues were picked up in the arts: John Osborne's *Look Back in Anger* dealt with the gulf between the working and middle classes; John Braine's *Room at the Top* dealt with class mobility; and Alan Sillitoe's *Saturday Night, Sunday Morning* dealt with the newly non-deferential working classes. In these cases class was joined by images of the north of England (establishing distinctions between north/south, urban/industrial and rural/non-industrial), notwithstanding the situation of London; in Nell Dunn's *Up the Junction* the robust optimism of the metropolitan working class was celebrated.

So what was involved in these class journeys? The person who made the journey had to leave behind one self and accept the construction of a new self. This could involve: (i) a new set of friends; (ii) a new set of colleagues; (iii) a new schedule of tastes in consumer goods; (iv) a new schedule of tastes in moral and aesthetic judgements; (v) a new place to live in a new part of town (along with new party political allegiances); and (vi) it would also involve a new voice and vocabulary. It all entails the adoption of a new form-of-life.[31]

The experience of change can also involve loss, and this figured centrally in the post-Second World War period: either the loss of the familiar status hierarchies of pre-war days or of the comfortable certainties of working class community. In the novel *Brideshead Revisited* Evelyn Waugh mourned the loss the pattern of life of the English landed aristocracy in the face of the plebeian world of the post-war modern welfare state: the former had great houses, servants and money whereas the latter had only mean banal dreams. The theme was reworked in Kazuo Ishiguro's *The Remains of the Day*. As it happens, the houses were rescued,[32] and the moneyed were fairly safe.[33] John LeCarre investigated this social world further: one can distinguish the early realistic novels from the later more stylized fictional worlds. In the former category, works such as *The Spy Who Came in From the Cold*, which examined the cold war, introduced themes of loss and betrayal. In the latter, starting with *Tinker, Tailor, Soldier, Spy*, we were introduced to George Smiley, the world of the circus and the reiterated themes of loss of empire and power, coupled with class change and political and personal betrayal.[34]

Journeys through popular culture

The nineteenth-century rise of industrial production generated the first mass markets. The general level of living rose throughout the nineteenth and twentieth centuries, and in the years following the Second World War economic recovery slowly gave rise to a 'consumer society'. In the early period the new goods were novel and rare – domestic electrical appliances, cars, maybe foreign holidays – later, the output of consumer goods was extensive. J. K. Galbraith[35] argued the system was absurd, with advertising creating the wants industry satisfied, thereby generating private affluence and public squalor. In the 1950s and 60s the disposable income of young adults plus the new opportunities of consumption gave rise to a distinctive sub-culture – 'youth culture'. The mix of youthful energy and commercial exploitation produced a series of fads and fashions, but sometimes the material had a wider impact. The Beatles taught a generation to make their own music and art rather than merely to import passively from the USA (or any other putatively authoritative source).[36] Also in this period Bob Dylan exemplified the involvement of popular music in political activity[37] and in the 1980s Bob Geldof moved pop stars into the territory of media-vehicled charity work. In all, a new sphere of cultural activity opened out: trashy, evanescent and occasionally influential.

The rapid economic advance of the post-Second World War period found more vigorous form in the 1980s, the time of economic boom associated with the government of Prime Minister Thatcher. This period was not merely one of economic advance but a celebration of consumption: as Gordon Gecko declared, 'greed is good'.[38] Postmodernists celebrated the end of meta-narratives in the achievement of a fabulously rich consumer society where one could assemble a life-style from proffered alternatives.[39] The idea had profound resonances: the political ideas of freedom were assimilated to consumer market-place choice and celebrated in media-carried equations of freedom with free choice.[40] Those who were left out – the poor – were understood as 'failed consumers' and subject to policing and control.[41]

In all this one can distinguish between popular culture, mass culture and everyday life. Popular culture is the popular sphere of industrial market society and mixes little tradition (implicitly authentic) with market-carried mass culture (implicitly something other than authentic). Mass culture is the realm of the market; inauthentic and manipulative. Everyday life looks to the ordinary routines of domestic and community living but reads these as embedded within a capitalist market society – for the individual and community benefits and disbenefits are intermingled.

The most influential line has been the cultural studies analysis of 'popular culture'. It has a distinct occasion. The invention of English Literature as an intellectually vigorous and ethically critical discipline belongs to the period after the Great War. It was the territory of high culture. The interest in English literature was a reaction to the slaughter of the Great War and the disillusionment in respect of the claims to civilization that had been made by Europeans.[42] In place of implausible general claims to progress, an avenue to moral progress could be identified more reliably in the guise of the lessons available within great literature. In this context, the work of F. R. Leavis and his colleagues offered an independent-minded radicalism which celebrated the moral worth of ordinary everyday life, and which could be accessed via the careful study of the canon of English literature. All this 1920s and 1930s work was the seedbed for the post-Second World War emergence of cultural studies.

International journeys – 'other cultures'

A related study of journeys is available which centres on the shift from one culture to another. It is concerned with Englishness as something encountered as a coherent given package and which has to be learned and judged according to the tenets of the received culture of the migrant or expatriate making the journey. It might take the form of recollections in adulthood of moves made in childhood, or more direct reports of moves made in adulthood.

The journeys of migrants, émigrés and expatriates

Benedict Anderson[43] has written about religious and colonial pilgrimages. Those who leave/arrive have an image of what they might become and they also have an idea of what they are leaving. Travel is not mere relocation in geographical space. It is bound up with ideas of self, the familiar and the as-yet unexpressed. A few examples can illustrate this familiar point.[44]

One can relocate internationally. In the nineteenth and twentieth centuries the British staffed their colonies and a group of colonial officials developed. Canadine[45] writes about 'ornamentalism' – the ways in which colonial officials and their families remade in the colonies the social worlds they had left behind, albeit a stylized version – a simulacrum – from which they looked back to home,[46] now another simulacrum.[47] In the post-war period many UK people emigrated to the white commonwealth. The migrant flow also ran from the former colonial periphery to the metropolitan core, famously aboard the *Empire Windrush* from the Caribbean to the UK. The majority were ordinarily skilled – middle class, working class – while some were artists, and their work drew on the journeys they had made. V. S. Naipaul travels from Trinidad, an island with a colonial history, where he was a member of an Indian minority amongst an African majority. Salman Rushdie travels from an Indian Muslim family to the secular world of the London literati – *Midnight's Children* and *The Satanic Verses* deal with the journeys, and in *Fury* the journey is into a middle-age in New York. Indeed, in recent years the increasingly internationalized global economy has allowed people to work overseas as 'expatriates'. It is an ambiguous status, in effect they are middle-class migrant workers where ordinarily such workers are poor, moving to work in richer countries.

One can relocate within a country. One can point to rural/urban migration: Dennis Potter contrasts his urban life with his Forest of Dean childhood in *The Singing Detective*; J. G. Ballard tracks his life from war-time Shanghai to the environs of Elstree in London in *Empire of the Sun* and *The Kindness of Women*. In a similar vein Paul Theroux in *The Kingdom by the Sea* records a journey around the coastline of Britain, offering an ethnography/commentary on contemporary England. Or one can relocate psychologically within local social networks: 'marrying out', thus seen as a problem for Jewish, Catholic and other religious groups; 'self-haters', a term used for Jews critical of Jewishness; and 'transferred nationalism', a term used by George Orwell for those who look to the model of other countries. J. G. Ballard, in *Millennium People*, also diagnoses the psychological withdrawal of the affluent in the form of 'the docile middle classes'.

There may be problems of re-entry. One can live and work outside one's home country, only to return and discover it is difficult to readapt. Jonathan Raban's novel *Foreign Land* details the re-entry problems of a long-term expatriate who returns to the UK only to discover it makes no sense and that the town he left overseas is home; the novel ends ambiguously, with the expatriate either returning to his (overseas) home or dying. Thus there may be problems of enduring psychological distance. Robert Chesshyre in *The Return of the Native Reporter* details his reactions to the UK having lived in the USA for four years. The UK is characterized as comprising a complacent incompetent elite; frustrated businessmen, grasping yuppies, a bitter working class and a detached under-class.

However, against the bitterness the journeys can be optimistic. In the post-war years the UK received many migrants – now around 8 per cent of the population – many from sometime colonial territories, and more recently from Central and Eastern Europe. It is true that there have been problems (thus racism and occasional riots), but there is also optimism. Monica Ali gets this beautifully in her story of Bangladeshi migrants. At the end of the novel the heroine is taken by her friends on a trip to the local ice-rink; it is the symbolic end-point of her journey:

> 'How are your boots, Amma?'.
> Nazneen turned around. To get on the ice physically – it hardly seemed to matter. in her mind she was already there.
> She said, 'But you can't skate in a sari.'
> Razia was already lacing her boots. 'This is England,' she said. 'You can do whatever you like.'[48]

Journeys through the stock of ideas/images

The material of history is contested. What counts as 'England' is not given. It is a matter of remembering and forgetting, not merely amongst professional historians or novelists but amongst politicians and nations. Patrick Wright speaks of the production of a 'national past' – an official history – a story of where we came from, who we are and by implication where we are going. The national past articulates a complex package. It is constructed and reconstructed: flags, parades and anthems, sacred sites, official truths (in particular, remembered war, a time when action made a difference and the country accumulated moral capital). A series of ideas and images of England can be sketched. All were mentioned in the earlier chapters, but now can be presented and summarized in plain view: rural, commercial, radical, multi-cultural and official.

A rural England: land, place and people

The rural version of England offers an organic image. It celebrates the long-established institutions – Church, law and parliament. It celebrates the concern for individuals-in-community. It celebrates the habit of modest consensual reform (exemplified in the English common law). It celebrates place, the love of country and countryside. It eschews nationalism in favour of patriotism, to characterize oneself as English is to point to place not nation.

Scruton[49] offers a view of England in which it can be modelled as an cyclical construction. A quartet of key institutions carry a great tradition (law, church, monarchy and parliament) which informs an official ideology and popular opinion (country, place and patriotism) which in turn informs civil society and little tradition (home/place, with clubs, societies and settlements). Scruton argues that loyalty to groups can be nurtured in more than one way: religion, dynastic authority or 'experience of home'.[50] England was understood as country, a home, a place,[51] not a doctrine,[52] not a state. The English were those who were born in the place, lived in the home.[53] The home was ordered by common law and was thus linked to land – the law of the land. The land/place/home idea endures whereas political regimes pass.[54] The common law affirms custom and tradition, so cases accumulate law. In this way the place was domesticated; it was also enchanted and ritualized. There were associations, uniforms and rules. The monarchy symbolized home,[55] deep attachment, personalized relations.[56] The English had a personal relationship to a place.[57] The institutions were made and they grew, ground-up. Church, parliament and law were associations of people: they allowed settlements and agreements, and thereafter people got on with their lives. Hierarchy made things clear.[58]

Scruton argues that in law we can regard organizations as persons. In social interaction there were many clubs and societies, and so England was full of corporate persons: 'And this is the way the English loved their country. England, for them, was a place of clubs and societies: it was a land saturated with a sense of membership'.[59] A country is build of layers: geography, climate, location, agriculture, society, culture, language and religion. A key organization was the Church – the Anglican settlement: 'England was a place … welded together … by an experience which was fundamentally religious in its meaning'.[60] The 'Anglican vision of England … [was] … an Arcadian landscape enchanted by laws and institutions, and made holy by ritual and prayer'.[61] A second organization was the common law, with its focus is on practicality. It accumulates material, it is judged discovered (rather than made). It is not European code. The focus is on individuals and routines and practice. It is an instrument for resolving problems, not social control.[62] Scruton argues the key to government is representation (not democracy). The English character was private, reserved and sceptical, traits which centred on place/home – they carried a pragmatic household ethic: '… while English believed in law and authority, they despised officialdom and distrusted the state.'[63] There were many civic organizations that represented interests and views, and they contributed to the whole. He adds that the system has been lately degraded as a secretive and manipulative state takes over. There has been a slide towards 'a secular republic, governed by conspiratorial elites, most of them resident elsewhere'.[64]

Finally, Scruton notes that the pastoral theme in English culture is central – gardens, country, but that interest now is in the idea of country life, not in actually living it.[65] Thus, '… sensing that they no long truly belong in the land which made them, they have lost their self-confidence as a people'.[66] The war, he says, generated 'moral fatigue'.[67] The country is being remade with urbanization/globalization. The virtues of England are fading.

One might say that Scruton gets the idea of locale but then reifies/mystifies it. It is not simply the place where we live. It is more – place is equated with land (it endures) and invested with mystery (it is enchanted). And the history – the rich detail of all the social processes running down through time – gets blurred into the story of a timeless enchanted place (reification/mystification). The timelessness is an aesthetic response, and the enchantment is religious. The aesthetic response to place is legitimate on its own (we all know places we think beautiful: Slad Valley, Batu Ferringhi, Victoria Harbour or the M42 in the winter rain with Bach on the car stereo), and the religious response is also legitimate – a place can be beautiful and hint at the divine – but Scruton hitches these aesthetic/religious responses to institutions – nation and Church. The former is approached crabwise – place and patriotism rather than nationalism (by implication, a mainland modern contrivance); the latter is the Anglican Church, presented as a key English institution.

Scruton offers an elegy to England.[68] It is a pleasure to read, but there are three crucial problems: (i) the aesthetic/religious responses are attached to particular institutions, but these are two different sorts of things – and the latter are not necessary for the former (art galleries and churches are not necessary to aesthetic or religious experiences); (ii) the responses and institutions mentioned are linked to nation, but once again there seem to be different sorts of things run together – the former is a strange aesthetico-institutional hybrid (as noted), while the latter is a way of grasping the totality of the ordered community within which one lives – the grasp does not have to be aesthetic/religious; and (iii) the aesthetico-religious approach to place/institution/nation issues is a construct which is time-less (fixed or given), but human life is lodged in time – the repetitive rhythms of days and years, the slowly unfolding patterns of family life, the wider dynamics of social change, the macro dynamics of complex change, and unfolding historical trajectories.[69]

A commercial England: innovative, entrepreneurial and expedient

The commercial vision is familiar, the tale points to energy and advance. The usual story speaks of the rise of commerce and empire, centring its descriptions on social history: the rise of towns, the development of industry and conflict with religion. It is an evolutionary tale of the construction of a commercial liberal-democratic society. Roy Porter offers a direct celebration,[70] looking at the role of England/Britain (not distinguished) in the making of the modern world. He points to the long eighteenth century. The Glorious Revolution of 1688 and the Act of Union of 1707 inaugurated a period of rapid development. The institutional settlement (the liberal parliament) offered a permissive environment in which diverse groups sought material advance. The eighteenth century, up to the confusions attending the French revolution, was one of unparallel success. Porter goes on to identify series of further crucial institutional vehicles: the realm of clubs, societies, associations and the new print media; civil society; the burgeoning commercial companies; and the natural scientific clubs and societies. These three are the keys to the advance, to what he calls the English Enlightenment. Their great tradition ideas encompassed materialism, empiricism, science, progress and happiness. Porter remarks: 'British pragmatism was more than mere worldliness: it embodied a *philosophy* of expediency, a dedication to the art, science and duty of living well in the here an now.'[71] In this context older institutions had to adjust: the Church slowly fell away; and landed property owners became improvers and joined the commercial realm, or they too fell away. Their little tradition was found in the social history of the development of the modern world, and the key to the vision was commerce. Commercial England/Britain comprised diverse strands: the early merchant traders; the early factory owners; the provincial magnates; and the city trading houses. The centre of gravity of commercial England was the southeast, London in particular, and commercial England is liberal. One might object to the term Enlightenment, as there is no modern nation-state, neither democracy nor a sovereign people. As Tom Nairn puts it, the liberal world was backward looking – the model was the Renaissance city-state or Dutch Staadtholderate. One might also object that the rural and urban poor do not appear in this work, nor do the subjects of overseas empire. Porter may be correct to celebrate the vigour of science and commerce but his history – and it is a familiar one – is as partial as that offered by celebrants of the rural vision of England.

A radical England: the common man, progress and democracy

The radical version of England looks to the energy of the masses, to the successive struggles waged in order to advance democracy and social welfare, the England of the Levellers, Tom Paine, the Chartists, the Independent Labour Party and so on. In the modern period the institutional vehicles of this vision of England can be found in the multiplicity of organizations established by the working classes, rural and urban; thus corresponding societies, trades unions, colleges, Church-related educational activities, newspapers and political parties. It is the territory explored by E. P. Thompson. The American and French revolutions inspired a domestic democratic movement, Chartism, but in the event it was defeated in the 1848 springtime of nations and thereafter the working classes turned towards ameliorist welfare within their communities. The process of political democratization was much slower, indeed ongoing. There have been various arguments for democratic and social reform, and the vision is usually of a species of republican democratic England. Their great tradition ideas centred on socialism (a series of diverse strands, as various programmes and agents are identified) and their little traditions encompassed the weapons of the weak – deference, aspiration and opposition. The key organization of radical politics came to be the Labour Party, much discussed, essentially subaltern conservative, devoted to slow reform. The post-war period saw significant success, built on the back of the work of Lord Beveridge, however, recent decades have seen a preference for liberal managerialism and the energy of the public sphere seems to be waning[72] as activists turn to new concerns and forms of engagement (the territory of single issue non-governmental organizations (NGOs), the internet, global protest movements and the like).

A multi-cultural England: empire, migration and diversity

The multi-cultural version of England is new, belonging to the post-Second World War period, when there was a large inward migration of former empire citizens from the West Indies, from South Asia and from Africa. These migrants were visible; their physiology marked them out. They were visible where other/earlier migrants were not. Their presence undermined received ideas that ran reductive arguments – grounding identity in race or ethnicity – and made it clear that identity was malleable; as the migrants adjusted to the UK so too did the UK adjust to them. A fraction of the host population objected but over time the majority welcomed the new diversity (symbolically, the UK's favourite food is now reported to be 'chicken tikka masala'). The experience of these relative newcomers (whose children and grandchildren are wholly local) argues that a diverse multi-cultural identity is not one to advocate or choose, but rather to acknowledge. It is how we are now. Its celebrants paradoxically often prefer to speak of Britain, thus 'British-Asian',[73] but it means that the hyphen has already begun to be used. The new context is Europe; more flows of people and more hyphens in prospect.

An Anglo-British official England

The official Anglo-British version of England has been the central area of concern for this text. The structural analysis made in the earlier chapters detailed the construction of the political-cultural project of Britain: the central institutional vehicles are those of the state, the project is top-down. The project included Scots, Welsh, Irish and English, but the latter were not especially enthusiastic about the union as England/Englishness seems to have been lost in Britain/Britishness. Yet the English were the majority population in the multi-national union of Britain. The elite affirmed the idea of Britain and the masses were encouraged to be obedient, a popular English nationalism was not encouraged. The idea of England continued, running alongside Britain, but it was an admixture: part residuum of an England long gone (the pre-British idea); part official memory (the 'olde merry England' confected in the early years of the nineteenth century as the Hanoverian monarchy settled in, finding antecedents for the kingdom in a line that ran back to medieval days thereby chopping out the massive involvement of France); and part the sets of ideas that run through little tradition; family, work and community. England became half-remembered, un-clear and only diffidently affirmed.[74]

Journeys into the political-cultural project of Europe

The era of 'block-time' offered a variant on received ideas of 'Britain/ Britishness' (discussed above) – a mix of the great and little traditions deployed within the confines of the American post-war project, and remarkably robust. Yet change now enfolds Britain and the British – internationalization, regionalization and the project of the EU. The elite will resist. For the masses the new European cultural space opens up the possibility of discarding Britishness in favour of a rediscovery of Englishness. A novel practice and discourse of Englishness could emerge. It would be a notion of Englishness in the context of Europe.

The post-war settlement came to an unexpected end in 1989/91. A new narrative is required. It is in the context of recent changes – the end of the 1989/91–2003 interregnum – that one might expect these debates to move forwards; stasis is not an option. As the patterns of life of the inhabitants of the Isles become ever more deeply intertwined with their neighbours on the mainland one would expect patterns of self-understanding to shift – the political-cultural project of Britain might now be in a process of eclipse. The key institutional vehicle for action and debate is the project for a united Europe. As an over-arching characterization, we can speak of the 'Europeanization' of the UK, however, this is clearly not going to happen in any simple way – it will be contested. One way to speculate about how the action and debate might unfold is to generate a series of 'scenarios'.

If one argues that structural change entails agent responses and that as elites read and react to enfolding circumstances they will formulate new projects, then it is inevitable that received ideas of Britishness/Englishness will change. It is not clear how they will change – shifting identities are the outcome of complex structural/political processes – but shifting identities also feed the processes of change – new ideas open up new lines of change. So how might the process of relocating England proceed? The relationship of the British state-regime and the developing project of the European Union is and will be contested – patterns of future events cannot be read-off structural changes –

none the less it is the case that the structural pressures for integration in Europe are very strong. One might sketch out three scenarios: first, confirm the project of Britain as part of the continuing American sphere; second, acknowledge years of political error and rally to the EU; and third, grant the long-term view that the inhabitants of the UK have 'nowhere else to go'.

In the first scenario, it has been argued that the political elite should relocate the project of Britain as an offshore part of the continuing American sphere, in rhetorical terms, Britain should join the North American Free Trade Agreement (NAFTA). Change has been slow, confused and reactionary. The preference of the anti-European conservatives is for continued links with the USA– one might identify two variants: the tacit and the explicit. The former line suggests that the interlinkages of Britain and the USA and EU are now very deep and any significant variation of these linkages would be either difficult or unnecessarily divisive (or both). In this case a modest sensible movement forwards with existing relationships is proposed: the EU continues to develop slowly, NATO remains as the cornerstone of European defence, and the general issue of EU/USA relations should continue to be fudged. Such a tacit strategy has the advantage of being non-dramatic, but it has the obvious disadvantage of ignoring the key problem – the relationship of the EU and the USA. It also has, many would argue, the insuperable problem of ignoring the evident commitments of France and Germany in particular to continuing European integration. The explicit version argues the case directly – it is said that membership of the EU was a mistake and should be renegotiated, and that the British should ally themselves with the USA. The Iraq war offered a variant in that US Secretary of Defense Donald Rumsfeld's 'new Europe' was seen as part of a wider alliance with the USA and the liberal market, business-oriented English-speaking/using countries.

The second scenario suggests that the British elite acknowledges the lessons of political error and rallies to the EU. It was suggested, for example, that Prime Minister Blair was a transformed politician domestically – no longer weak and vacillating, rather strong and determined and able to over-rule doubters in his party and the village and call a referendum on the euro. This was the preference of distraught Euro-enthusiasts, but it was implausible: polls recorded a hardening of popular opinion against the euro (following the Iraq crisis split with the mainland), key figures in the New Labour party were hostile, the villagers remain hostile, and Blair was a weak prime minister.

The final scenario takes a long-term view, arguing that the British oligarchy have 'nowhere else to go', and that structural circumstances will oblige an accommodation to the project of Europe. On the one hand there are deepening linkages with the mainland, while on the other American foreign policy is realist, hence the US foreign policy community is likely to pay attention to the powerful mainland core of Europe, rather than a semi-detached periphery. The question becomes, therefore, what will such an accommodation look like and how might it interact with the structurally occasioned possibility of relocating England?

Change, identity and action

Change is routine. Periods of stability can be contrasted with episodes of relatively rapid change. The nature of the UK polity and its relationship to Europe is now directly in question. The project of the European Union offers a new political-cultural territory. It is a space in the process of being shaped, and is within the political-cultural space of the developing European Union that a notion of England might be relocated.

Notes

1. A position argued for in earlier chapters – see Peter Winch 1958 (2nd edn 1990), *The Idea of a Social Science and its Relation to Philosophy*, London, Routledge, and Alasdair MacIntyre 1985, *After Virtue: A Study in Moral Theory*, London, Duckworth.
2. In regard to the elite, a position argued by H. Young 1999, *This Blessed Plot: Britain and Europe from Churchill to Blair*, London, Macmillan.
3. An argument taken from E. Gellner 1964, *Thought and Change*, London, Weidenfeld.
4. The implications of the latter for the former could be problematical as the official British version of the Second World War would have to be revised in order to grant, for example, that the German people were both aggressors and victims, and that British moral heroism also included terror bombing German cities and their civilian populations.
5. The route into this issue is narrow, a review of available sets of ideas. Another route would be via practice, an investigation of just what was going on. After Jacques Darras – we could consider the lives and ideas of the ordinary men and women – truckers, holiday makers, students, second homers, retirees, Elizabeth David fans and so on – who are making Europe, but this is for another research.
6. Clifford Geertz, cited in Fred Inglis 1993, *Cultural Studies*, Oxford, Blackwell.
7. Nor is it an area where I would claim great expertise – once again these remarks are relatively untutored – they are also focused on the English scene.
8. My expertise is slight – again, the material is needed – and my focus will again be on the English scene, thereafter English-language material. The work of French, German and Swiss theorists is not dealt with directly.
9. Chris Jenks (1993, *Culture*, London, Routledge) reviews the multiplicity of meanings of the term 'culture' as it appears over the years within the traditions of social science. The uses and origins of the term within philosophy and literature are

noted: (i) the nineteenth-century reaction to industrialism which stressed culture; (ii) the classical notion of culture as civilization in opposition to barbarism; (iii) the sociological and anthropological notions of culture as sets of ideas or, more broadly, social practices; (iv) the German philosophical notion of culture as acquired learning/sensibility; and (v) a contemporary Anglo-American usage within social sciences which looks to the diversity of social practices of groups and sub-groups. All of which is distilled into four meanings: (i) culture as a cognitive category, a state of mind (with an ideal goal implied) [arts and ethics]; (ii) culture as a level of collective social development [culture and society]; (iii) culture as the arts [arts]; and (iv) culture as the way of life of a people [praxis]. See also Raymond Williams 1980, *Keywords*, London, Fontana.

10 After Z. Bauman (1973, *Culture as Praxis*, London, Routledge) we could instance (i) base/superstructure – culture as the reflected realm of ideas; (ii) acquired skills – culture as learning – becoming a cultured person – this overlaps with status; (iii) total world view – culture as the total set of ideas evident in a society ('British culture' or 'American culture' or 'modern culture'); (iv) culture as praxis – the sets of ideas carried in routine social practice (form-of-life – language game).

11 Raymond Williams (1958, *Culture and Society*, Harmondsworth, Penguin) details optimists and pessimists in a UK context, noting the benefits and costs of industrialization. S. Hughes (1979, *Consciousness and Society: The Reorientation of European Social Thought 1890–1930*, Brighton, Harvester) considers some reactions to modernity – in the UK one pervasive theme is a nostalgic ruralism, for example Edwin Lutyens' cottages, William Morris's arts and crafts movement, Ebeneezer Howard's garden cities, the inter-war preference for suburbs, not cities. Such rural nostalgia runs down to the present, as can be seen with the new 'suburb villages', the Campaign for Rural England, and the suggestion that farmers should lose CAP subsidies but be paid for maintaining 'the countryside'. See also R. Scruton (2001, *England: An Elegy*, London, Pimlico) on England as home/place, and on the UK movements, see R. Colls 2002, *Identities of England*, Oxford University Press.

12 See Jenks 1983; Colls 2002.

13 See Inglis 1993.

14 In this text the notion of an interpretive-critical social science has been affirmed. One might speak of modes of social theoretic engagement – ways of drawing down on received tradition in order to make arguments for particular audiences. The formal statements of social scientists reflect the intermingled demands and concerns of context, theorist and audience – the theorist is not neutral and the demands/concerns which shape social scientific work can be accessed in reflection, which is crucial. One might say that reflexive self-embedding is a necessary condition of scholarship – the theorist must locate himself within the flow of social processes within which interventions are made. The idea of journey privileges certain strategies of reflexivity: it allows an element of biography; it allows an element of comparison; it allows an element of judgement. Clifford Geertz (cited in Inglis 1993) sees culture as the ensemble of stories which we tell about ourselves, and suggests that biography – the story we tell about our-self – is therefore a way of accessing the whole cultural ensemble (biography/social world intermingle). C. Wright Mills (1959 (1970) *The Sociological Imagination*, Harmondsworth, Penguin) speaks of the sociological imagination picking up where private concerns and public issues overlap – in this way we can address both private and public concerns in a disciplined rational fashion (biography/social world/scholarship intermingle). And Edward Said (in his Reith Lecture) argued that scholarship does not side with the powerful, rather with the weak – it is critical. Analysts can interrogate their own trajectories, offer comparisons of past and present (or old places and new places) and offer judgements of the present.

15 And these remarks are largely untutored. The particular concern here is with literature, English-language writing and commentary, more specifically that which concerns itself with change and identity. It is a restricted borrowing.

16 The standard disciplines affirm particular ideas about identity (schematically): economics/politics affirm liberal ideas – autonomous, coherent and knowledgeable selves ('rational'); whereas sociology/cultural studies more inclined to look at sets of relationships ('constructivist'). See P. W. Preston 1997, *Political-Cultural Identity: Citizens and Nations in a Global Era*, London, Sage.

17 The idea comes in varieties, some, paradoxically, rather manipulative: thus Ervin Goffman (we present self); others, curiously self-effacing, thus postmodernists (self dissolves in non-serious expressive consumption, we are how we shop). The idea is also available outside Western culture: thus John Clammer on Japan, selves are social but remain concerned with sincerity/achievement. See John Clammer 1995, *Difference and Modernity: Social Theory and Contemporary Japanese Society*, London, Kegan Paul International.

18 Hannah Arendt in P. Wright 1985, *On Living in an Old Country*, London, Verso.

19 Thus one's mother tongue is a socially learned skill – it is not one that one has to think about – but as the UK modulates into Europe, English-language monolingualism looks outmoded.

20 See Preston 1997.

21 See K. Ishiguro 1987, *An Artist of the Floating World*, London, Faber and Faber; see also E. Waugh 1962, *Brideshead Revisted*, Harmondsworth, Penguin.

22 I take this from E. Gellner 1964, *Thought and Change*, London, Wiedenfeld.

23 So we mark rhetorically the compatibility (not incompatibility) of the claims made here with those made earlier by using the same trio of ideas – locale, network and memory – to order enquiry into journeys.

24 R. Williams 1958, *Culture and Society*, Harmondsworth, Penguin.

25 A later version is available from Bea Campbell, recreating Orwell's journey through English class to Wigan (1984, *Wigan Pier Revisited*, London, Virago).

26 P. Townsend 1979, *Poverty in the United Kingdom*, University of California Press.

27 His magnum opus: E. P. Thompson 1968, *The Making of the English Working Class*, Harmondsworth, Penguin.

28 K. Kumar (2003, *The Making of English National Identity*, Cambridge University Press) speaks of a 'moment of Englishness' (pp. 202–7), that is, English nationalism. Citing Paul Langford he locates the idea of Englishness as early nineteenth century (1805), but argues that the mix of medievalism, Anglo-Saxonism and ideas of 'olde England' only became popular in the late nineteenth century as

Britain's pre-eminence waned. Kumar argues that this pragmatic, rural, Anglo-Saxon idea remains fairly central to the English self-identification – it is the conservative line.

29 P. Wright 1985; also P. Wright 1993, *A Journey Through the Ruins*, London, Flamingo; P. Wright 1995, *The Village that Died for England: The Strange Story of Tyneham*, London, Jonathan Cape.

30 One satire on a progressive attempt to rationalize class mobility was offered by M. Young (1958, *The Rise of the Meritocracy*, London, Thames and Hudson). He speculated on the evasions likely to be made by middle-class parents confronted with the possibility of their children moving down the hierarchy. It might be noted that nineteenth-century novels were full of this anxiety – highborn people fallen on hard times, the struggles of genteel poverty.

31 R. Hoggart 1989, *A Local Habitation (Life and Times Vol 1: 1918–1940)*, Oxford University Press, p. 156.

32 On this see Wright 1985 and the tale of the political/policy re-orientation of the National Trust: from land for workers to use recreationally to houses celebrating a vanishing form of life – the invention of 'heritage'.

33 Patterns of income and wealth distribution in the UK have not significantly altered over the post-Second World War period.

34 The theme of upper-class betrayal – the Cambridge spies – is now an established one within reflection upon the nature of the UK – part cliché (available to lazy writers), part running sore.

35 J. K. Galbraith 1958, *The Affluent Society*, Harmondsworth, Penguin.

36 An observation taken from Hanif Kureishi.

37 M. Marquese 2003, *Chimes of Freedom: The Politics of Bob Dylan's Art*, London, The New Press.

38 A character in the film *Wall Street.*

39 For explorations of consumption see the journal *Theory, Culture and Society.*

40 F. Jameson 1991, *Postmodernism: Or the Cultural Logic of Late Capitalism*, London, Verso.

41 Z. Bauman 1988, *Freedom*, Open University Press.

42 Jenks 1983.

43 B. Anderson 1983, *Imagined Communities*, London, Verso.

44 Anecdotally, my own experience of boarding long-haul jets: bound by received ideas in the departure hall, free of them once seated in the aircraft, shifting from ascribed to achieved status (to the extent that the stereotypes of others allow) and escaping thereby the stifling demands of the British class system.

45 D. Canadine 2001, *Ornamentalism: How the British Saw their Empire*, London, Allen Lane.

46 The overseas contingent of the polity made an impact on its overall – including domestic – definition. The idea of 'British' emerged in the North American colonies where there was a diaspora – a 'British world'. See S. Schama 2003, *A History of Britain III: The Fate of Empire 1776–2001*, London, BBC; C. Bridge and K. Fedorowich (eds) 2003, *The British World: Diaspora, Culture and Identity*, London, Frank Cass.

47 R. Scruton 2001, *England: An Elegy*, London, Pimlico.
48 M. Ali 2003, *Brick Lane*, London: Doubleday, p. 413.
49 Scruton 2001.
50 Scruton 2001, p. 5.
51 Scruton 2001, pp. 2–3.
52 The preference is for patriotism, not nationalism. J. C. D. Clark (2003, *Our Shadowed Present: Modernism, Postmodernism and History*, London, Atlantic Books) argues that a common identity came from 'civilization, law and history' (p. 88) rather than English nationalism or British nationalism. The idea of patriotism is picked up by R. Weight 2002, *Patriots: National Identity in Britain 1940–2000*, London, Pan. It seems to me that the idea of patriotism is a disingenuous nationalism or an unreflexive nationalism – a refusal to reflect on the occasion of nation-mystification/reification.
53 Scruton 2001, p. 7 Scruton also remarks that 'home' was made clear via empire – overseas one remembered home (p. 2).
54 Scruton 2001, p. 9.
55 Scruton 2001, p. 12.
56 Scruton 2001, p. 13.
57 Scruton 2001, p. 15.
58 Scruton 2001, p. 21.
59 Scruton 2001, p. 72.
60 Scruton 2001, p. 85.
61 Scruton 2001, p. 111.
62 Scruton 2001, chapter 6, Scruton argues that English society centred on common-sense and on law and that the usual tales told about class are false – classes are groups of cooperation. See also Canadine 1988 – class is association, not structural, and conflict is neither central nor inevitable nor particularly visible in history.
63 Scruton 2001, p. 57.
64 Scruton 2001, p. 198.
65 Scruton 2001, p. 242.
66 Scruton 2001, p. 243.
67 Scruton 2001, p. 249.
68 One comparison: Peter Ackroyd's *London: The Biography* (2000, London, Chatto) is a similar celebration of place – not the changing patterns of life of people (as with Patrick Wright on Dalston Lane), but place as meaning-drenched or past-drenched. It is the same reification and mystification. Wright speaks of 'auratic sites' – the place which is particularly exemplificatory of the ideas of the community. Scruton wants the whole of rural England to be auratic, just as Ackroyd's London is auratic.
69 See also P. Hitchens 2000, *The Abolition of Britain*, London, Quartet. This is a demotic version of Scruton, providing similar before/after comparisons. Interestingly, Hitchens gets the nature of the Second World War correct, reporting it as a disaster which resulted in friendly occupation. He now asks if the British can recover their nerve, although to do what, other than oppose the EU, is not clear. One further note – these conservative commentators have difficulty with

Britain/England, thus Clark 2003 views Britain as a phase in the longer history of England.

70 R. Porter 2000, *Enlightenment: Britain and the Creation of the Modern World*, London, Allen Lane.

71 Porter 2000, p. 15.

72 On this see D. Marquand 2004, *New Statesman* 19 January. Marquand's book was not available to me at the time this text went to press, but see D. Marquand 2004, *Decline of the Public*, Cambridge, Polity.

73 See for example Y. Alibai-Brown (2001, *Who Do We Think We Are: Imagining the New Britain*, Harmondsworth Penguin; also Sharma 2003, pp. 418–22) who points to the difficulties of settling, to the problems of racism, and to the problems of cultural self-identification. That the term 'British Asian' has come to be used is odd – it points back to an empire that conquered and enslaved their forebears. One might ask what is wrong with 'English-Asian'. See Zadie Smith 2001, *White Teeth*, Harmondsworth Penguin; Hanif Kureishi 1990, *The Buddha of Suburbia*, London, Faber and Faber; Monica Ali 2003.

74 In Julian Barnes' *England, England* (1998, London, Picador) the whole country/identity is relocated to the Isle of White where it is recreated as a theme park – a popular simulacrum – which slowly becomes the reality: a country feeding upon its own cultural history, that is, going nowhere.

10 Storytelling II: discourses of Europe

The idea of Europe is contested.[1] A number of discourses can be identified;[2] or, cast in substantive terms, there are a number of 'Europes', each a particular historical configuration, each writing its own history.[3] The most recent of these discourses/configurations was 'block-time'[4] – 'state-socialism' versus 'the free world'[5] – but the 1989/91 end of the cold war undermined such familiar identities, and the project of the European Union came to the fore. As we have seen, it has its own history, structures and trajectory. It is a nascent polity, yet its nature is also contested. It is in the process of re-constitution: thus, the expansion to the East, the aspirations for a new constitution and the differing reactions to the Iraq war. The process of determining a new direction (or restating the old) requires a fresh collective exercise in 'remembering and forgetting'.[6] It is in this exercise that a new idea of Europe will be found and constructed.

The political-cultural identity of contemporary Europe is a matter for debate[7] – issues of the history, the present day and the likely futures of the continent. The debate is fluid. Social analysts, including historians, write for particular audiences, and these, more often than not, have been national in character. Such analyses have formed one aspect of the production of 'national pasts', however, a European audience is in prospect, requiring a 'European past'. In this novel context, a series of themes might be identified:[8] (i) aspects of collapse (war, occupation and memory); (ii) aspects of recovery (planning, affluence and renaissance); and (iii) aspects of unification. It is through the process of running these intermingled, dense, shifting sets of debates, that a new discourse of Europe, a new or restated project, might be constituted.

Aspects of collapse I: Europe and the experience of war

The process of the shift to the modern world in Europe was violent (as it has been elsewhere[9]): dynastic empires disappeared, and new nation-states were brought into existence (some to be snuffed out again, or reabsorbed within wider empire-like systems[10]). In the twentieth century the violence affected whole peoples, it was not restricted to elites or soldiers (plus those non-combatants unfortunate enough to be in the wrong place), rather its reach extended throughout society: in the Great War, indirectly (casualty lists, later, war memorials); in the Second World War, directly (invasions, population displacements and the bombing of towns and civilians[11]). The violence has had major economic, social, cultural and political consequences. It is part of the modern experience of Europe and Europeans.

The early part of the twentieth century saw intermittent wars in Europe. Though these were viewed at the time as discrete wars between nation-states, the period is now amenable to re-labelling as an interlinked series of conflicts, together constituting a thirty-one-year-long civil war (1914–45). Two mottoes encapsulate matters: Norman Davies's[12] bitter joke, the title of a table in the annex of a book – 'Europe's wars: a selection' – which recalls that the continent has had *so very many* wars; and Robert Kee's[13] phrase – 'the world we left behind' – which points to 1939 as a dividing line in European historical experience or, put another way, that with the Second World War, Europeans finally contrived a war so destructive that it led to the occupation, division and systematic reconstruction of their polities.

The Great War – its impacts, political and cultural (the war and the early 1920s)

The shift to the modern world began in Europe. A variety of aspects could be considered: ideas, (religion, philosophy and the rise of science), episodes (the rise of towns), activities (the growth of commerce), exchanges (the start of 'voyages of discovery'), plunder (the flow of credit-creating bullion from Spanish Latin America), or accidents (the way this complex of factors led to the invention of 'capitalism'). Over the period 1400–1700 (say), Europeans produced a very dynamic form-of-life:[14] both domestically, with intensification (innovatory, rationalizing (efficiency), exploitative (unequal) and productive); and internationally, with expansion (the exchange with other cultures, trade, absorption and reconstruction[15]).

The form-of-life was successful in its own terms (and ours); celebrated in the idea of progress (a contested idea then, and more so now). The self-understanding of the form-of-life developed down the decades, with great traditions (elite ideas in politics, the arts, literature, religion and the like) and little traditions (the judgements of ordinary life, popular culture and media). One aspect of self-understanding affirmed the notion of civilization, the achievements of the Europeans (in contrast to all others, variously deficient). There was a widespread cultural confidence. The domestic advance, geographical expansion and celebration of the form-of-life continued unabated through to the early years of the twentieth century.

The Great War had multiple impacts upon the empires and nations involved. In retrospect, it signalled the start of a process of elaborate cultural self-destruction, as the continent engineered its own eclipse. At the time, however, the confidence carried over into war fighting: it was going to be 'all over by Christmas'. The claims to cultural success fell away as the continent slipped into military stalemate. In Britain, the early optimism of domestic mobilization was followed by military incompetence (the trenches, myths of lions led by donkeys and needless sacrifice) and there were later unfulfilled promises of 'homes fit for heroes'. In France there was an early optimism, culminating in the battle (myth) of Verdun, and later widespread defeatism. In Germany too an early optimism gave way to endurance until late in 1918 when there was an abrupt collapse in the war effort. The regime fell, generating the myth of the 'stab in the back'. More broadly, a series of venerable dynastic empires collapsed and new nation-states made their appearance. But Europe was no longer the dynamic centre of global capitalism (notwithstanding the 1920s economic boom). European political life was disturbed, with domestic upheavals, and new centres of international power. European society was disturbed.[16] There were expectations of radical change (in the event, mostly dashed). The self-understandings of Europeans as civilized, advanced and so on, were sharply undermined.

At this distance, these events, both familiar (in memory) and distant (as history), and their contemporary resonances are matters to revisit in the EU context. The Great War involved a series of conflicts and a number of participants, each with their own memories. This opens a new area of enquiry : a European-level history. A European-level memory would supplement or replace the multiplicity of available national and group memories. A series of (deliberately naïve) questions present themselves: (i) how did the elites manage to make the mistake of going to war? (ii) what was the point at which reality dawned? and thereafter (iii) how did people grasp the scale of cultural and social damage?

The revolution of 1917 and the short twentieth century

As the Russians made their revolution it meant that one combatant dropped out of the Great War and a major European dynasty collapsed. Eric

Hobsbawm's notion of the 'short twentieth century' signals the importance of the experience. The USSR was a positive example for many social reformers, but a negative example for conservatives. There was a series of widening circles of reaction to the revolution: elite shock in Europe and extensive foreign[17] anti-communist intervention in the Russian civil war. In the 1920s, 30s and 40s the revolution offered both encouragement and problems for the European left: the USSR was either an example or a warning, hence the social democrat versus communist split (an issue that ran down into the post-Second World War period). In a rather different fashion the example of the USSR fed into critiques of the liberal economic system and contributed to the rise of preference for planning.[18] A series of theorists in the UK argued the case for planning,[19] as did the neo-institutionalist theorists of the New Deal in the USA.[20] In France the popular front made similar efforts, and in Germany there was analogous activity by state: both in the Weimar period and with the national socialist state.[21] The influence of the revolution ran even wider, however. The Soviet regime was anti-colonial and generated support and sympathy in the colonies of the Europeans, Americans and Japanese. In Europe there was an hostile domestic reaction from the established elites: in the UK there were attacks on the labour movement and widespread elite sympathy for fascism; in France, there was extensive political conflict; in Germany, left revolutions were defeated, contributing to the rise of fascism; and the USA saw the use of red-baiting. The reaction found further expression in the cold war and the equation of socialism and fascism in the rhetoric of totalitarianism, matters only a recent part of Europe's past: all issues to be debated rather than forgotten.[22]

The 1920s and 1930s – the experience of fascism

The political-cultural project of fascism came in varieties (primarily European, but also East Asian[23]): a southern European Catholic clerico-fascism (Franco, Caetano, Mussolini, deVallera); a northern European version (in Germany); and numerous other groups throughout Europe (including, of course, Sir Oswald Mosley and the British Union of Fascists[24]). The European fascist movements were hostile to the available pattern of modernity (rational, democratic, liberal-market atomistic, etc.[25]), but a pool of alternative ideas was available – conservatism, romanticism, mysticism, militarism, racism – and provided the means to a celebration of hierarchy, community and action.

The history of European fascism is brief. The core trio of countries were Germany, Italy and Spain, and in each case the fascist movement came to power utilizing extra-parliamentary violence, the first victims of which were their domestic populations. The reactions in other European countries varied.

The memory of the Great War inhibited direct responses, and there were many 'fellow travellers of the right'.[26] In the event, the strategy of 'appeasement' was adopted.[27] The slide towards war gathered pace and the final disaster began to unfold with the 1938 *Anschluss* which saw Austria absorbed within the Third Reich.[28]

A series of social scientific explanations of European fascism have been offered:

(i) psychological – mad people caused it all;[29]
(ii) international relations – the unjust end to Great War, exemplified in the Versailles Treaty, generated a fascist reaction;[30]
(iii) political-historical – the circumstances of interwar Europe saw mass poverty and insecurity, and fascism emerged as the creed of the resentful provincial lower middle classes, both tolerated and encouraged by conservative elites – once the project was up and running it developed its own logic and ran out of control;[31]
(iv) political economic – fascism was an extreme variant of capitalism;
(v) cultural historical – fascism was an extreme expression of modernity.[32]

One might point to a tale that looks to unfolding contingent dynamics of change,[33] borrowing from the available strands of commentary, pointing in particular to the legacies of war[34] and the effects of depression[35] coupled with elite incompetence[36] and mass powerlessness.[37] The history of European fascism presents two key issues to revisit in an EU context: the extent to which the history is acknowledged in contemporary public debate (selective memory and ritualized reference[38]); and whether emphatic nationalist or fascist-style ideas are a continuing element of European social thought.[39]

The experience of the Second World War

As we have previously mentioned, the middle period of the twentieth century saw Europeans involved in a series of major wars: different wars, involving different people, lasting different lengths of time and remembered differently. The Second World War, itself a series of intermingled sub-wars,[40] was a catastrophe for Europeans. Thereafter, there were further wars marking the end of empire: in East Asia (Dutch East Indies, Malaya and French Indo-China); in the Middle East (Palestine, Egypt (Suez)); and in North Africa (Algeria). After 1945 the cold war quickly developed, identifying Europe (East and West) as a potential battleground in a war that could become nuclear, thereby producing widespread anxiety in regard to an eventuality that would have reduced the continent to radioactive ash.

The sequence of wars resulted in great loss of life, destruction of accumulated infrastructure (public and private property) and social dislocation. The

Second World War resulted in systematic political and social change: hence Robert Kee speaking of 'the world we left behind'. In September 1939 the Europeans had worldwide empires; by May 1945 the continent was ruined, divided, occupied and eclipsed. The impact of war had been catastrophic, with destruction and death on a scale that defeats the contemporary imagination. The events of the Second World War shaped subsequent European history and they shape European memory, both official and popular. The first broad expression is the idea of the 'free West' (its Eastern European counterpart being 'state socialism'); its more recent expression, in process of construction, centres on the EU.[41]

A number of histories are available, addressed to national audiences, inflected by the demands of the official truths of block-time. They are selective; thus the British national past identifies an heroic military/moral victory, and omits the catastrophe, loss of empire and subsumption within the US sphere. The scale of Europe's wars, in retrospect, is quite extraordinary. The intermingled sequence, in the context of the developing EU polity, presents a series of issues for a contemporary analysis. In respect of war: (i) the systematic killing (by whom, of whom and justified how) – industrialized warfare (pursued directly, machine guns, barbed wire, gas, tanks, aeroplanes, and pursued on base of the mass mobilization of entire national populations) – the bureaucratic mass production of death – the civilian deaths – the camps[42] – the bombed cities (from Guernica to Dresden[43]); (ii) the political incompetence (how did the elites manage to generate the catastrophe?); and (iii) the (continuing) extent of martial traditions.[44]

Aspects of collapse II: Europe – remembering and forgetting, 1945–48

In the context of the preparation of a 'European history of Europe' established national pasts will have to be amended. The re-imagining will be a broad social process (not a matter for authoritative elite discussion and decision). A process of multiple re-imaginings can be anticipated: like the EU itself, these are likely to be jumbled and contested rather than neat and tidy. It is difficult to see any analogy at EU level of the sort of authoritatively secured political-cultural closure needed to establish a single national past (or single European past). It is also difficult to see how Europeans can escape the intellectual and moral imperative to write the history openly and well. One starting point will be the constructions and confections of the 1945–48 period.

The Second World War radically changed the patterns of life of Europeans. The episode had political implications, with pre-war regimes swept away and new settlements made throughout Europe. The continent was occupied and divided. The USA and the USSR were the key players and novel high political projects were affirmed. More profoundly, there was a pervasive cultural impact: the ways in which Europeans thought of themselves, their societies and histories, plus, by implication, prospective futures, were all called into question. The war years weakened available European 'national pasts',[45] as the stories available in routine practice before the war began, the various banal nationalisms,[46] were undermined: the British lost their empire; the French had been invaded; the Germans saw their country collapse in defeat and so on throughout the rest of Europe in a series of cultural catastrophes. A series of exercises in 'remembering and forgetting' had to be undertaken: a new set of stories established, a new set of self-images adequate to a shattered present, able to deal with a catastrophic recent past and effective in sketching out a liveable and believable future. In all, it was an elaborate process of re-constructing political-cultural identities.

In late 1945 and 1946, in the chaos of contemporary Europe, the population faced a choice: run seminars to discover what had happened and who was to blame, or begin clearing the rubble. The choice for the latter was inevitable, but it entailed the rapid construction of an agreed understanding of what had happened. Tony Judt[47] has looked at the process of establishing an official/popular memory of the Second World War. In the American occupied part of Europe the process of making 'the West' began. Blame for the war was attached to the Germans – more particularly, the national socialist regime – they were purged (partially), the myth of resistance grew and a route to the future could be imagined. In the Soviet block a similar process produced a rather different agreed understanding: the war was read as a struggle against fascism. The defeated national socialists were blamed and the new socialist system was begun, and again a route to the future was imagined. Judt argues that much of the official/common memory, East and West, was poor history. The end of the war saw a spread of local civil wars develop and many people and groups accommodated themselves to events. Yet the business of rebuilding demanded some political-cultural settlement (the more awkward as there was neither a peace treaty process nor harmony, and the cold war soon began to build). These matters remain unclear and unsettled. However, the remembering and forgetting served to establish foundation myths for contemporary Europe (and the EU).

In the realm of official memory there were: (i) surrender agreements between Germany and the Allies, agreements between the Allies about occupation zones within Germany, agreements between the Allies about spheres of military responsibility within Europe, and implied agreements about spheres of influence within Europe for great powers; (ii) the task of re-establishing the state system (Germany as aggressor, others as victims); and (iii) a series of official surveys, histories and tribunals. On the subtle matter of official forgetting, passive (not worth remembering) and active (where forgetting is preferred[48]), the behaviour of all involved was screened, some matters were pursued, others were discretely set aside.

First, in respect of the *behaviour of the axis powers* ('things that were not done to us, but done to others whom we did not care to remember'), there were hierarchies of suffering. (i) The camps: the scale was noted but not much pursued, and the survivors went about their business, not embracing a status as survivor; the idea of the holocaust came much later.[49] (ii) The other victims: the slave labourers, the displaced persons (those who had fled the fighting), the expellees (those who had been expelled from the areas where they lived, for example, Sudeten Germans and others in Eastern Europe[50]). (iii) The first victims: the German population (which included a series of groups – trades unionists, opposition politicians, church figures, the mentally handicapped and of course the German Jewish population). (iv) Those the Allies found useful: German research scientists, members of the German security services (as the cold war began), members of the local populations generally (the businessmen, the local low-level officials and so on).

Then, second, the *behaviour of the allied powers* ('things which we did to them'). (i) Asymmetrical justice: principles applied to them but not to us – the defendants at Nuremberg were prohibited from calling attention to Allied behaviour in their defences. (ii) Asymmetrical justice: Allied behaviour was not examined according to the criteria being used to judge the defendants (both individual and symbolic), thus the British and American bombing of German cities (ideas of 'area bombing' and 'destroying enemy morale' are euphemisms for systematically killing German civilians). (iii) Selective justice: some guilty were prosecuted but many were not, it would have been not merely time-consuming but also inconvenient to try everyone (the German scientists etc.). (iv) The extensive nature of the accommodation, collaboration and support for the national socialist authorities during the periods when European countries had been occupied. And finally, in respect of *both (all) sides*, the dehumanization and racism: German claims to 'master-race' status; the British wartime hatred of Germans; the Russian hatred of Germans; the American hatred of the Japanese; American internal race divisions; the Japanese hatred of Americans, ditto Chinese, ditto Koreans and so on.[51]

The sphere of popular memory flowed from personal experience: in the services, in fighting or in civilian life at home. Popular memory was inflected by the claims of the official realm: sacrifice, patriotism, moral victory and the promise of future social advance. The actual experiences of countries differed in degree, but all began with the damage of the war years.

At the present time, looking at an idea of 'Europe', these experiences must be revisited. The USA was not damaged by war and emerged powerful and prosperous. Indeed, the USA and USSR – which suffered twenty million dead

– shaped the post-Second World War world. State socialism emerged in eastern parts of Europe and the 'Allied scheme of history'[52] (or 'the West'[53]) in western parts. However, the impacts went deeper, shaping the self-understanding of Europeans, and the lessons continue to unfold (though the learning is uneven).

Europe and occupation: 1945–53 ('Uncles Sam and Joe')

At the elite political level the interests of the 'big four' – the USA, USSR, UK and France – were different: most immediately in running the military occupation of Germany, and thereafter in setting the shape of the post-war world. The USA and USSR were dominant in what became their spheres. Allies had to fall in line. The immediate post-war period 1945/46 had specific difficulties: dislocation, displaced people and the business of setting up governments and beginning the task of rebuilding. It was a confused period but decisions were taken that shaped post-war Europe.

The wartime Allies ran down different trajectories, they had different experiences and understandings, and these fed into the debates about the shape of the post-war world. The view of the American political elite emerged at Yalta and Potsdam in their intention to establish a liberal trading sphere. Having emerged from the war undamaged and powerful, and as they occupied the less damaged part of the continent (Western Europe remained relatively unscathed up until 1944[54]), the US had extensive plans for the shape of the post-war world.[55] In contrast, the Soviet elite read the war in terms of the defeat of fascism in the Great Patriotic War. The situation in the Soviet-controlled areas was difficult as the war in the East had been extremely destructive.[56] The politics of big four's relationships during the immediate post-war period saw the wartime alliance begin to fail. In the UK, the 1945/46 election of the Labour Party signalled a desire for change but shortages continued amongst the problems of organizing rebuilding. In Germany 1945/46 was year zero – the women cleared the rubble and rebuilding began. In France de Gaulle and the Free French returned and there came the first acknowledgement of the social costs of occupation with thereafter many problems of reconstruction (as there were throughout the continent). In America the completion of the Pacific War in August 1945 lead to demobilization in an untouched country. In contrast with the pre-war depression this was the start of the building of a consumer society, the start of the post-war American dream. In the USSR meanwhile, they faced the legacy of extreme destruction, with twenty million dead. After victory in the Great Patriotic War however, they emerged as a superpower – but with shallow local support in Eastern Europe, within a few years they were ruling over an autarchic sphere.

The cold war, the division of Europe, occupation and 'block-time'

After Raymond Aron,[57] one can speak the cold war as the discipline/legitimation of foreign occupation. The block-time settlement was contingent, subject to change, internal criticism and decay and external events and shocks. The occupation lasted from 1945–89/91 and set trio of tasks for the elites: establishing, sustaining and adjusting block-truths.

The task of establishing block-truths began amongst the elites. One might point to the politics. There were, for example, sharp debates between the US and the UK on the nature of Bretton Woods system. The European elites did not spontaneously embrace US pre-eminence, though the US was in a position to insist. Many new institutions were established, all of which carried expectations – some were successful, others not. The task of establishing block-truths amongst the populations entailed the dissemination of official truths – 'the Allied scheme of history' – and demobilizing proponents of counter-truths by capturing, demobilizing or undercutting their institutional bases (buying newspapers, starting others, undermining unions, encouraging others etc.). Once the block-truth was up and running it had to be sustained.[58] As an official truth it developed its own status and momentum. It became a political-cultural paradigm. It set the core ideas, established their boundaries, policed them, and put them to work in reading events, in Europe and elsewhere.[59] As an official truth, it became familiar and easy – a ritual. Everyone affirms official truths not because they are true but because they help to secure social order.[60]

Over the period 1989/91 the Soviet occupation of Eastern Europe ended and the USSR was dissolved. It was the start of a sequence of changes. The Americans and Europeans reacted with both celebration and dismay. In 1989/91 the American occupation of Western Europe was still in place, familiar, amiable and seemingly permanent. One could argue that occupation continued after 1989/91 with a somewhat reworked (or redundant) NATO coupled with eastward expansion. But the overall rationale had gone: the official block-truth that had legitimated occupation was no longer tenable. The cold war ended when the official enemy dissolved away. The political-cultural paradigm was overthrown by events. Yet the official truth was implausibly sustained. Despite the interregnum[61] ('we won' or 'the end of history' or 'globalization'), it must have been clear to the elite (as to the mass) that the familiar game was over. The first signal of the construction and recognition of a new Western European political-cultural paradigm was the focus on the European Union. The project of the EU became central to European politics, though as it was re-animated in Maastricht in 1992 there were internal tensions in Europe and US hostility. Finally, the Bush regime's Iraq war provided the catalyst for open debate, the real start of the post-cold war period and the end of the occupation in Western Europe.

Life in the block-spheres: state socialism/the free world

The years following the end of the Second World War saw a sharp reorientation of political life for European elites as domestic agendas were circumscribed by the impacts of the war years (destruction, dislocation, social upheaval etc.), the reform demands of their populations and the over-riding imperatives of the new political-cultural paradigm of block-time. In the Soviet sphere there was a slow recovery from the catastrophe of war. Eastern Europe had seen the greatest destruction yet the Marshall Plan was constructed so as to rule out post-war aid from the undamaged USA. An autarchic Eastern block was established. The death of Stalin and the subsequent government of Krushchev allowed some reforms, and the long era of Brezhnev provided modest comfort, plus the burdens of military competition. The lack of domestic reform, the unceasing burdens of military expenditure and the rise of images of Western consumerism eventually undermined the legitimacy of the system.

In contrast to the autarchic Eastern block, the USA established a large open trading area. The economic life of the West was centred on the cities of Washington and New York. The IMF, IBRD and the economics ministries of the US government were in Washington. The global financial centre of Wall Street was in New York, so too the UN headquarters. It was to these cities that European leaders made ritualized pilgrimage journeys,[62] for meetings, photo opportunities and to meet the international media. Davies[63] identifies a series of elements in this 'Allied scheme of history': (i) belief in a uniquely valuable secular Western civilization (the pinnacle is the 'Atlantic community'); (ii) the ideology of anti-fascism and the enduring moral credit derived from its defeat; (iii) demonological fascination with Germany, the centre of responsibility; (iv) a romanticized view of Russia as an ally; (v) an acceptance of the division of Europe; and (vi) 'The studied neglect of all facts which do not add credence to the above'.[64] The confection fed into politics, scholarship[65] and popular opinion.

The masses of Western European countries had a double experience – prosperity and a novel cultural example. The wartime destruction was extensive, yet the recovery was rapid and these countries entered what came to be called 'the long boom' – a lengthy period of unprecedented economic prosperity. The USA become a cultural example; it was seen as the future – optimistic, progressive and free from the hidebound traditions that suffused Europe (there were significant flows of migrants into the US). The positive image lasted through the 1950s and only began to fade as US problems mounted in the 1960s.

There are matters to revisit, issues for a contemporary analysis. The broad European view has inflected and shaped national histories, but a European history might now be anticipated, and its shape will flow from debate. The experiences of the USSR and the USA marked Europe. The end of the block-time variant of the former Eastern Europe – state socialism – was read as a return to Europe. A decade or so later, however, perceptions are shifting – in Central Europe there are doubts about Brussels and in Russia there are signs of recovery (plus some nostalgia for Soviet days). In Western Europe, if we think of historical trajectories, the unfolding contingent patterns of change running down through time, the processes whose accumulated cultural residues underlie the arrangements of the present, then we must acknowledge that Europe's present is shaped by the experience of some fifty years of American occupation. These legacies have to be detailed. In particular in these matters, clarity is a necessary condition of rebalancing the relationship of the USA and Europe.

Aspects of recovery I: unexpected, dramatic and novel, 1947–62

The commitment to reconstruction was widespread and found different expressions. In the UK, an elite concern to re-establish the country's position came to be expressed in terms of the intersection of three spheres, the USA, Europe and Commonwealth, plus there was a strong popular commitment to securing social progress with the establishment of the welfare state. In Germany the elite faced de-Nazification, the resumption of sovereignty and the moral and political re-establishment of the country as well as reconstruction, where corporatist traditions continued in the social market economy. In France the elite sought recovery from the occupation and collaboration and the moral and political re-establishment of the country, and confronted the need for reconstruction. More broadly, there was widespread interest in co-operation.

There was a series of national political cultural projects – we can detail the ways in which elite projects intersected with mass expectations, the ways in which elite projects intersected, and recall the scale and character of the success. The Europe we inhabit today took recognizable shape over this period – a mix of domestic successes, early moves towards cooperation and the overriding discipline of block-time.

Reconstruction and the role of planning, 1947–57

In the 1940s the idea of planning was up and running. Debates had been provoked by the emergence of the USSR. There had been debates in 1930s in regard to the perceived failure of free-market liberalism (in USA, UK and fascist-ruled countries). There were debates about decolonization and development. There was experience of war planning. The Marshall Plan required recipient governments to plan the use of monies received.

The scale of the European disaster is the start point. In the 1930s there was a series of wars. All involved European countries, directly or indirectly, but the scale of the destruction in the Second World War was immense: millions of soldiers killed or wounded; millions of civilians killed or wounded; enormous financial and economic costs of war materiel; the costs of destroyed and damaged infrastructure (houses, industrial units, roads, railways, canals, bridges, water/sewerage, electricity/gas, schools, hospitals, government buildings and so on); and there were millions of displaced persons, refugees and expellees. The level of European economies in 1945 was reduced to that of 1900: the tasks of recovery were great, in the event so too were the rates of recovery.

The idea of planning was widely popular, with debates provoked by emergence of USSR, the perceived failure of free-market liberalism and the task of decolonization and development. Planning systems were put to use both during the war and after – wartime planning for production was extensive in the USA, UK,[66] Germany and Russia and a series of new techniques of management were developed. After the war planning was required in Western Europe by Marshall Plan authorities and similarly instigated in Eastern Europe by the Soviet Union. In Western Europe planning fed into the construction of post-war European mixed economies or social democratic systems. In eastern Europe planning was an integral feature of the system. Plan optimism fed into the ECSC (the background is the functionalism of the interwar period which looked to role of experts running functional units in place of nation-states), and the idea was popular with the block-leaders: the USA organized the Bretton Woods system (economic institution building, vehicles for planning – instruments both for reconstructing Western Europe and for making the American-cantered post-Second World War liberal trading sphere); and the Soviet Union exported state socialist planning systems and created the integrated Council for Mutual Economic Co-operation (Comecon) economic block.

All this feeds through Monnet/Schuman into the ECSC, Euratom and the Rome Treaty. The historical development of the EU can be read in terms of the projects of the founding fathers and the development of relevant institutional mechanisms. It has grown as a series of accumulations following elite decisions, slow developments (in economics, society, law etc.) or abrupt events (demanding a response). A series of phases can be identified: an early optimistic phase (the ECSC, the Rome Treaty of 1957, an early period of success); a second pessimistic phase, through the 1970s (Euro-sclerosis but with accumulation of members and functions); and a recent burst of activity from the middle 1980s (the 1985 SEA and the 1992 TEU, a period of rapid advance); and finally (arguably), the debates of the post-2003 war/crisis period.

In the Anglo-American official discourses of the 1980s and 90s the claims of planning were decisively rejected in favour of 'market solutions', however we can recall the experience of planning – both the confidence in it, and its fall from grace: (i) early American scepticism; (ii) early New Right organizations influenced by Friedrich Hayek, Karl Popper and Milton Friedman, such as the Montperlin Society; and (iii) the triumph of the New Right in the 1980s. The debate about states and markets is ongoing. The liberal talk of spontaneous order is absurd, as markets are subtle social constructs; however, the scope of rational planning is an open question.[67]

Le Plan – the Monnet Plan 1947–52 (and onwards)

In France the elite faced the tasks of recovery from the occupation/collaboration and the moral and political re-establishment of the country. The task of economic reconstruction was urgent. The politics involved looking to a future rapprochement with Germany. Three wars in under a century made this task urgent. French political policy looked to a Europeanized future, and opposition to the USA and the UK (seen as the former's creature) flowed from de Gaulle's experience in the war years (when he was opposed by Roosevelt). The economic policy associated with Jean Monnet, who correctly diagnosed the deep-seated problems of modernization needed by the French economy and began an indicative planning system, is widely seen as successful.[68]

Wirtschaftswunder – 1948 currency reform onwards

In Germany the elite faced de-Nazification, the task of the resumption of sovereignty and the moral and political re-establishment of the country. The reconstruction of the war-ravaged country was also central. The key political figure was Adenauer, who was chancellor for many years. The new capital was in the Rhineland city of Bonn and commentators spoke of 'the Bonn republic'. It was a success. The key economic figure was Ehrhardt; his currency reform of 1948 established the Deutschmark and also separated the three Western zones from the Soviet zone. The policy thereafter affirmed a 'social market system', once again a success.

The Festival of Britain – victorious war and the welfare state

In mainland Europe, following the material devastation and occupation (read after the event as ethico-political and cultural devastation), any simple return to the *status quo ante* was not possible – reconstruction and recovery had to have a prospective element. Yet in the UK, where continuity was key, a great myth was constructed: 'the virtuous victorious war.' This myth had problems: militarily it was misleading – the UK survived in 1940/41 and thereafter was the junior partner in the stronger military alliance (the USA and USSR); diplomatically it was misleading – the UK was a great power in 1939 but by 1945 (or shortly thereafter) it was a subsidiary element of the new US-dominated liberal trading sphere; economically it was misleading – the UK was centre of a powerful empire/economic block in 1939, by 1945 this was gone and the UK depended upon American loans; culturally/morally (internally) it was misleading – the country was exhausted and divided and determined upon social change (the welfare state). Culturally/morally (externally), however, the claim carried more weight, as in mainland Europe the UK was seen as one of the 'victorious Allies'.

The myth served to disguise the extent of the changes in the international and domestic situation of the UK and to encourage an attachment (both among the elite and the masses) to a version of the pre-war model of the UK – a low-level unconscious conservatism (later expressed as 'heritage') acting as a brake on more radical change (thus, symptomatically, the UK did not join the EEC). The achievement of the Atlee government was modest ameliorist politics, an exercise in overdue social welfare, the construction of the welfare state.

The UK elite begins the long exercise in post-Second World War consolation (or a refusal of 'ordinary-ness'[69]): the Commonwealth (in place of empire); Greece to America's Rome (in place of a diminished autonomy); the new Elizabethan age (in place of post-Second World War problems); and the Festival of Britain (in place of acknowledgement of loss). The policy concern to re-establish the status of the UK came to be expressed in terms of locating the country at the intersection of three spheres – the USA, Europe and the Commonwealth. In the event, the elite were absorbed into the American sphere, the masses were politically demobilized, content with welfare, and a long debate about 'decline' began.

The EU project

As we have noted earlier, the project began soon after the war. Its inauguration involved a mix of optimism (the key agents on the mainland), luck (the elite fractions in power favoured advance) and disdain (the response of the British).

The project accumulated effective power over the decades: never neat and tidy, but always moving forwards towards 'ever closer union'. The project was a part of the politics of the 1950s – the ECSC, Euratom and the EEC – though it became much more important over the next decade.[70]

All of which offers much to revisit. However, one key issue in the developing European context might be the nature and current viability of the European social model.

Aspects of recovery II: affluence, conflict and renaissance, 1962–89

In 1962 the Cuban missile crisis dominated Western high politics. In the same year the Beatles released their first singles. The prosperity of the 1950s extended into the new decade; the affluence was self-evident, so too the conflict. The renaissance is more debateable: there was a recovery of cultural confidence, a sweeping social liberalization and the rise of an optimistic European political left. All these were intellectual and cultural movements with many strands: political, social scientific, literary, film and so on.

Economic success, social change and political breaks (abrupt reconfigurations)

The post-Second World War 'long boom' (which ran until the oil shocks, debt and stagflation of the early 1970s) created conditions of unparallel economic affluence.[71] Mass consumer capitalism saw new patterns of consumption amongst middle and working classes. It saw social changes. It saw domestic demands for political and social reform as the post-war settlement fractured (in other words the political elites were slow to adjust to economic and social change, but political changes when they came were abrupt). The scale of trouble was relatively minor, but significant: Adenauer gave way; de Gaulle gave way; and Macmillan gave way. The more dramatic trouble engulfed the block-leader: the civil rights movement, the student movement and the Vietnam War meant that claims to political and moral pre-eminence for the USA were weakened.

The post-Second World War settlement in Europe revolved around the role of the USA. The Bretton Woods system governed economic activity whilst the rhetoric of cold war plus the apparatus of NATO (and similar organizations elsewhere – the Central Treaty Organization (CENTO), the Southeast Asia Treaty Organization (SEATO)) governed the politics (the cold war provided an over-arching official truth – the business of free market, free West and freedom). The system was successful for some twenty years or so, a period of stability and prosperity in the heartlands of the West. However, it began to unravel in the 1960s. The tale is familiar: it begins in the USA with the rise of the civil rights movement, the Vietnam War, draft resistance amongst young men and the growth of an oppositional culture/student movement.[72] These novel social and political phenomena undermined available understandings of the polity. What had been seen as a fundamentally stable functionally integrated society, a model for all other societies to aspire to or learn from, suddenly looked conflict-ridden. In place of the always complacently misleading and now overtly ridiculous view a plethora of new debates sprang into life. One downstream consequence was a renewed interest amongst Europeans in their own intellectual traditions.

In Western Europe there were analogous events/debates. In West Germany the reconstruction was successful. The government was confronted with a vigorous student movement, calling for democratization (involving necessarily critiques of post-war West Germany[73]). Intermixed with these groups there was an extra-parliamentary left – groups having recourse to insurrectionary/terrorist strategies designed to reveal the inherent violence of the capitalist system (the Red Army Faction). In Italy there were similar problems. An entrenched Christian Democratic elite, with links to the mafia, the Church and America was challenged by a vigorous communist party (the idea of 'Euro-communism'), the student movement and an extra-parliamentary left. There was also an extra-parliamentary right, representing a continuation of the pre-war fascist tradition. In France the problems were again repeated. The France of de Gaulle was challenged in the 1960s by the communist party espousing 'Euro-communism' and also by a student movement calling for reform and democratization. The process reaches a climax in the spring of 1968 when France came to the edge of revolution. In the UK there were some demonstrations – for example at Hornsea and Guilford Schools of Art, disputes at the London School of Economics, the anti-Vietnam War demonstrations – but the more significant changes were a schedule of agreed social reforms (laws governing censorship and sexual mores).

The pattern was repeated in Eastern Europe. The rule of Stalin gave way to Khrushchev in 1953. There were reforms and also failures. Khrushchev was replaced by Brezhnev in 1964, who oversaw a long period of slow economic advance that slowly turned into stagnation. Among other Eastern European countries there had been a series of reactions to Soviet rule, including attempts at reform. In Czechoslovakia the 'Prague Spring' (1968) initiated by Alexander Dubczek was an open attempt at liberal reform, suppressed in a Soviet-inspired Warsaw pact intervention/invasion. The Soviet block entered a slow decline, with the final collapse ordered by presidents Gorbachev and Yeltsin.

Brussels and the European project – the 1985 SEA

As noted, through this period, the political-cultural project of Europe continued to advance. The progress of the organization was intermittent and attended by disputes, however in the 1980s an old aspiration to economic integration was re-invigorated, championed by two politicians – Roy Jenkins and then Jacques Delors. The culmination of these discussions was the 1985 Single European Act – a plan to complete the single market – and the European project was reanimated.

Overall, once again the detail is interesting but there are wider issues to pursue: the nature of the European social model and the continuing value of the classical European tradition of social theorizing.

Aspects of unification I: after the cold war, 1989/91–2003

The West reached its apogee as a political-cultural project in the 1980s, seemingly successful in all regards. At that moment, its defining counterpart, the Soviet Union withdrew from cold war conflict and then collapsed. As the dust settled, the bi-polar world of American and European elite political imagination gave way to the ambiguities of a nascent tri-polar world; East Asia emerged as an economic and political power, and so too did Europe.

The end of the cold war in Europe – the end of block-time

In the USSR Leonid Brezhnev oversaw a long period of economic success follwed by stagnation, which continued after the Prague Spring of 1968. Subsequent leaders Andropov and Chernenko offered no solutions, but Gorbachev came to power in 1985 and inaugurated reform programmes – *glasnost* and *perestroika*. Over the summer of 1989 East Germans made their way via Czechoslovakia and Hungary into Austria, and thence to Germany, and in the autumn civic movements in East Germany took to the streets, holding massive demonstrations calling for reform. In Germany, Erich Honeker asked for their military suppression, but Gorbachev refused and the East German regime collapsed. In November the Berlin Wall opened. A ripple of further changes ran through Eastern Europe – communist parties stepped aside and liberal-democratic-style elections were held, with new governments winning power. The reform programme continued in the USSR until it was interrupted by a coup attempt; Gorbachev was replaced by Yeltsin and the USSR dissolved into confusion, Numerous new states appeared, Russia underwent a process of 'third-worldization' and signs of stabilization took a decade to appear.

Impacts and responses

The cold war had many aspects: political, military and economic competition, and its end provoked a series of responses, schematically: (i) celebration – the view that the West had won and the ethico-political end of history had arrived; (ii) regret – at the loss of clarity of block-time arrangements; and (iii) tension – the view that the end of the cold war had coincided with the rise of a new geo-economic context (familiar conflicts over trade between the USA and Japan were added by unfamiliar conflicts over trade between the USA and the EU).

A key element had been the provision of an over-arching block-truth. As the cold war faded, so too did the plausibility of the available official truths. Judt[74] noted an interlude, a period where agreed over-arching political truths were absent, debate unfocused ('morbid symptoms'). The debates moved towards a collapse, if not unexpected in general, then certainly unexpected in its occasion (arguments about aggressive warfare), timing (out of nowhere, or if not that, then in line with the electoral demands of us politics, that is, President Bush's mid-term election) and form (not amicable, nor even polite, but angry and hostile).

In the work that appeared it seems there was a retrospectively identifiable anxiety as to the continued viability of the project of 'the West': Fukuyama celebrated liberal-democracy (ignoring distinctions between the USA and the EU); Huntington also affirmed the continued importance of 'the West' in face of ethico-religious cultural competitors; and Bobbit, the latest in the line, speaks of the long war, where imperialism gave way to competition between fascism, communism and parliamentary democracy, the latter winning in the globalization-inflected guise of market democracy. All are affirmations of the model of the West/USA. The response within Europe was equally clear: Maastricht 1992 TEU; expansion 1995 Austria, Finland and Sweden; Amsterdam Treaty 1999, Nice Treaty 2000 and further institutional reforms; further expansion to East – Czech Republic, Poland, Hungary, Slovenia, Estonia and Cyprus; and in the future a possible further wave.[75]

The election of President Bush, the events of 9/11 and the Iraq crisis marked a decisive parting of the ways. The Iraq situation was confected inside the Washington beltway: President Bush was captured by a coterie of American nationalists, the mid-term elections had to be won and thereafter the proposed second Gulf War took on its own momentum. The debates surrounding these issues and events reveal a deep divide between the EU and the USA: what was latent during the 1989/91–2003 interregnum is now in the open and what has been said cannot now be un-said. The debate will have to be run until a new post-cold war consensus (contested compromise) is secured. In this sense, President Bush has been good for the EU. A return to the *status quo ante*, the cold war hegemony of the USA within Europe, is not possible. It seems that buried anxieties have broken the surface: an explicit discussion about the future of the EU plus an explicit discussion about the relationship of the EU and the USA.

Europe today: one, many or none?

One can identify a series of conjunctures and discourses of Europe (using a mix of historical structural analysis to identify and locate discrete phases, coupled with political-cultural analysis unpacking the sets of ideas which animated various elite and mass actors). In these discussions a series of issue complexes have been identified. They illuminate aspects of Europe's history. They illuminate aspects of Europe's present. They allow us to unpack discrete phases in the contemporary history of Europe, each transmitting residues down to the next (memories, institutions and projects for the future).

One can finally ask what is the character of the contemporary conjuncture and discourse of Europe, the one we inhabit. Cold war block-time is finished and US block-leadership is redundant (the implications of this continue to be at issue – as elite debates unfold it is variously recognized, resisted or acted upon). The key institutional centre invoked in discussions of Europe has been the European Union. The Iraq crisis shows divergences of view between European elites (less so European publics): the UK elite's continuing commitment to 'dual parasitism' – on the USA for political leadership and on the EU as a loose free trade area; the French elite's commitment to a European Union independent of the USA; the German elite's break with the US and preference for a federal democratic European Union (necessarily independent). Other elites offered different views. The widespread (Western) European public opposition to the US role in the Iraq war reveals a similar distancing from US and preference for European ideas (in favour of law and multi-lateralism and unenthusiastic about wars of aggression). We need to grasp the contemporary discourse(s) of Europe – elite-level projects, popular lines of opinion – and we might need to distinguish superficial rhetoric from deeper political stances.

The EU has developed over some fifty-odd years. Laffan *et al.*[76] offer a way of thinking about this which escapes received theorizing: an actor-oriented approach reveals the multiplicity of players and their manifold interactions and shows how over time these build interlocking sets of institutions. The focus is on social processes: over time these accumulate institutions, law and ideals. It is all contingent. The structures of the EU are ad hoc, interlocking/overlapping and robust (actors have learned to deal with this situation). One might point to political cultural identity in the same way, or discourses of contemporary Europe. The key is routine social practice. A series of routes to the future might be speculatively identified – one, many or none.

One – a single Europe – as an objective it seems implausible over the long run and impossible over the short run (too much elite disorder and popular scepticism). As a report on actual practice it does work, but only in a restricted area, as it calls attention to the activities of Brussels – there is an EU, it is up an running and it does do a lot (i.e. economic activity and law) but equally clearly it is limited in scope at the present time

Many – there might be many Europes – as an objective it is one for anti-Europeans, many Europes implies either a utopian return to the pre-civil war *status quo ante* (the Europe of great powers and empires of 1914) or a continuation of nominally independent nation-states subordinate to the political and economic leadership of the USA. As a report on extant social processes the idea is much more plausible: there are many communities within Europe, they have subtly different forms-of-life (the EU now reaches from the north cape to Sicily, and from Ireland's Atlantic coast to the plains of Central Europe – 400-odd million people), they read and react to the project of EU in diverse ways. In terms of individual identity we could start to acknowledge this, plus the intermixing that is developing as economies integrate, by using hyphens – thus 'English-European', and so on.

None – there will be no Europe – as an objective it is for the European refuseniks, and in the UK these are right-wingers who wish they were American, while in other parts of Europe there are other refuseniks – nationalists – who prefer the *status quo* or some particular *status quo ante*. As a report on social practice, it is difficult to see how this can be made; there is European intermixing and separating out the national strands would take an EU version of disintegration of Yugoslavia.

One might finally conclude by recalling that the idea of Europe is contested. It is a banal enough remark, yet it shifts us into the territory of process and detail. The question of how Europe will come to be understood will unfold over time. It is a matter of the understandings developed and lodged in routine practice. It is not a matter of elite definition. Over time, a clearer image or ideal is likely to emerge, but not soon, and not smoothly. The same might be expected in the case of England. However, there is one difference: the Europeans will discover or invent themselves with or without the presence of the inhabitants of the archipelago, the Isles, but the English can only restate their identity in the context of Europe. Outside the European Union all that would seem to be on offer, the only route to the future, is some variant of the present: on the part of the elite, a sad, futile aspirant-Americanism, obedient, biddable and always hopeful of acknowledgment and reward; and for the masses, disengagement, consumption and the rich consolations of family and friends.[77]

Notes

1 T. Christiansen, K. E. Jorgensen and A. Weiner (eds) 2001, *The Social Construction of Europe*, London, Sage.
2 G. Delanty 1995, *Inventing Europe: Ideas, Identity and Reality*, London, Macmillan.
3 The image comes from M. Foucault (obviously). The world we inhabit is fluid, not open for us to remake. We dwell within epochs, conjunctures or phases and the histories prepared are bound by context. 'Phase-centrism' is the analogue of every other 'centrism' and the cure is the same – reflexivity. See Delanty 1995 on the standard Greece-derived tale. Jacques Derras (J. Darras and D. Snowman 1990, *Beyond the Tunnel of History*, London, Macmillan) locates modern Europe's birth in the trading cities of the early modern period in northern Europe – all the familiar talk about Greece is a later confection.
4 An expression which originates, so far as I am aware, with E. P. Thompson.
5 N. Davies (1997, *Europe: A History*, London, Pimlico, pp. 39–42) speaks of the 'Allied scheme of history' as one way of unpacking the confection.
6 T. Judt 2002, 'The Past is Another Country: Myth and Memory in Post-war Europe' in J. W. Muller (ed.), *Memory and Power in Post War Europe*, Cambridge University Press.
7 There is no essence – the popular tale of evolutionary ascent from a start point in Ancient Greece is merely one tale (Darras and Snowman 1990). Nor can an essence be identified by committee – Brussels cannot do the job (Cris Shore 2000, *Building Europe: The Cultural Politics of European Integration*, London, Routledge), although it might be added that history does give examples of 'empire top-down' identities such as Britain, Russia and India. See B. Anderson 1983, *Imagined Communities*, London, Verso.
8 It is my list, flowing from my readings of contemporary scholarship and public commentary. The treatment here is 'broad brush' – others might set up the debate differently – but it is a debate (after Judt 2002) that will have to be run.
9 See B. Moore 1966, *The Social Origins of Dictatorship and Democracy*, London, Allen Lane.
10 Davies 1997 lists the European polities that didn't make it – that is, they emerged and then disappeared. In the twentieth century the political map of Europe shifted and changed several times – see annex III, p. 1268.
11 One might sum this from Guernica to Dresden (Pablo Picasso's painting, to Kurt Vonnegut's novel – K. Vonnegut 1970, *Slaughterhouse Five*, London, Jonathan Cape).
12 Davies 1997.
13 R. Kee 1984, *1939 The World We Left Behind*, London, Weidenfeld.
14 See for example E. Gellner 1991, *Plough, Sword and Book*, London, Paladin; E. Gellner 1992, *Reason and Culture*, Oxford, Blackwell. This is the key, of course, to all Marx's work.
15 P. Worsley 1984, *The Three Worlds: Culture and World Development*, London, Weidenfeld.

16 See Arno Mayer 1981, *The Persistence of the Old Regime*, New York, Croom Helm; E. H. Carr (1939) 2001, *The Twenty Years Crisis*, London, Palgrave.
17 Thus: British, French, American and Japanese.
18 See F. Clairmonte 1960, *Economic Liberalism and Underdevelopment*, Bombay, Asia Publishing House.
19 Keynes, Carr, Manheim, Macmillan, Beveridge *et al.* On the influence of the idea, see P. Addison 1977, *The Road to 1945*, London, Jonathan Cape.
20 See P. W. Preston 1981, *Development Theory*, London, Routledge.
21 See M. Mazower 1998, *Dark Continent: Europe's Twentieth Century*, New York, Alfred Knopf.
22 See Judt 2002.
23 Thus, the KMT (Nationalist Party) in China. The ideas were also current in Japan.
24 On Britain's Oswald Mosley and the wider spread of sympathisers see R. Griffiths 1980, *Fellow Travellers of the Right*, Oxford University Press.
25 Mazower 1998.
26 Griffiths 1980.
27 Not irrationally, though few would now defend Prime Minister Neville Chamberlin – see E. H. Carr 2001.
28 A series of dates could be specified: March 1938, *Anschluss*; September 1938, Munich; and March 1939, Czechoslovakia seized.
29 For a discussion of this theory in the recent war in Yugoslavia see M. Ignatieff 1994, *Blood and Belonging: Journeys into the New Nationalism*, London, Vintage.
30 A line with distinguished proponents such as J. M. Keynes, but see also M. Macmillan 2003, *Peacemakers: Six Months that Changed the World*, London, John Murray.
31 On the issue of the mass killing of Jews, see A. Mayer 1988, *Why Did the Heavens Not Darken: The 'Final Solution' in History*, New York, Pantheon. He argues that there was no holocaust, but instead a series of hatreds of Jews. The systematic killing was the scapegoating of a minority in a losing war of unparallelled barbarism – that is, they were in the wrong place at the wrong time.
32 See Z. Bauman 1989, *Modernity and the Holocaust*, Cambridge, Polity.
33 See Mayer 1988.
34 J. M. Keynes warned against the reparations burden placed on Germany, and the collapse of multi-ethnic empires was read in terms of Wilsonian liberalism, hence nations should inhabit states, which is fine but created a widespread problem of minorities. Also the League of Nations did not provide a system of security.
35 There was widespread economic and social distress, which Hobsbawm argues was generated by the USA who had financial power without any appreciation of responsibilities, and whose orthodox economic remedies made things worse.
36 On all the debates about 'appeasement' in regard to Germany see W. L. Shirer 1960, *The Rise and Fall of the Third Reich*, London, Secker and Warburg.
37 In Germany, Italy and Spain fascist domestic success suppressed local opposition; in France and Britain there was elite anxiety and sympathy; in Eastern Europe there was a multiplicity of local tensions – see Davies 1997.

38 The extent of fascism is underplayed (or, in the UK, where Mosley is treated as a comedy figure, virtually denied) or ritualized – the holocaust industry. The episode is reduced, neatly boxed and removed from European history (a measure of complicity between European elites and Zionist activists).
39 Ignatieff 1994.
40 See Davies 1997, Judt 2002.
41 Davies 1997, pp. 39–46.
42 Bauman 1989.
43 This is the European experience, but there is a wider history of air warfare starting in the 1920s (thus Harris attacks the Iraqis). There are also other war theatres – thus Americans attacking Japanese cities, ending in Hiroshima and Nagasaki, matters they cannot discuss in public (the Smithsonian scandal). On these bombings, see B. Cummings 1999, *Parallax Visions*, Duke University Press. The Japanese attacked cities throughout China and East Asia. The issue is slowly being considered in Germany – see W. G. Sebald 2004, *On the Natural History of Destruction*, Harmondsworth, Penguin.
44 One theme in recent commentary points to the intrinsic relationship of Britain and violence, a French elite concern for greatness, and contrasts the widespread revulsion to war in Germany.
45 Patrick Wright 1985, *On Living in an Old Country*, London, Verso. The 'national past' – the largely unselfconscious stories we tell ourselves about who we are, where we came from and where we are going – binds us into the political community we inhabit, itself, in the modern world, lodged within a nation-state. It binds us to the nation and reconciles us subjectively to the state – it is intensely political and it is contested.
46 M. Billig 1995, *Banal Nationalism*, London, Sage.
47 Judt 2002.
48 The theme of John LeCarre 1960, *A Small Town in Germany*, London, Pan.
49 N. Finkelstein 2001, *The Holocaust Industry*, London, Verso.
50 G. Grass 2002, *Crabwalk*, London, Faber.
51 See C. Thorne 1986, *The Far Eastern War: States and Societies 1941–45*, London, Counterpoint.
52 Davies 1997, pp. 39–40.
53 J. C. D. Clark 2003, *Our Shadowed Present: Modernism, Postmodernism and History*, London, Atlantic Books, chapter 7. Clark argues that the idea of 'the West' is an uneasy American confection, part of their post-war project. It excludes Germany, reading the country as having 'gone wrong'. This can be compared with the same tale told of Japan – in both cases history is subordinated to the demands of post-war US-sponsored reconstruction.
54 Judt 2002.
55 G. Kolko 1968, *The Politics of War*, New York, Vintage. The former has been said to lead to an 'American myth of Yalta' – that all had agreed to a liberal-democratic future for all Europe; the second saw Truman disinclined to compromise having new atomic weapons and thus military pre-eminence.
56 Mazower 1998; Mayer 1988.

57 R. Aron 1973, *The Imperial Republic: The US and the World 1945–73*, London, Weidenfeld.
58 See P. Berger and T. Luckmann (1966, *The Social Construction of Reality*, Harmondsworth, Penguin) on the 'machineries of universe maintenance'.
59 Bruce Cummings argues that the US mis-read the situation in Korea, where they saw nationalists as communists. The same goes for Vietnam and Cuba. Once the USA 'labels' them a self-reinforcing process is set in motion.
60 Official truths can decay, for example 'World Holocaust Day' where the remembrance of the Jewish victims of national socialism had degenerated into a banal public occasion attended by politicians, other 'worthies' and the media.
61 See Judt 2002.
62 See Anderson 1983.
63 Davies 1997, p. 40.
64 Davies 1997.
65 See Clark 2002, chapter 7.
66 P. Hennessy 1992, *Never Again: Britain 1945–51*, London, Jonathan Cape.
67 See P. W. Preston 1994, *Discourses of Development: State, Market and Polity in the Analysis of Complex Change*, Aldershot, Avebury; P. W. Preston 1996, *Pacific Asia in the Global System*, Oxford, Blackwell.
68 P. Calvocoressi 1997, *Fall Out: World War II and the Shaping of Postwar Europe*, London, Longman.
69 See T. Nairn 2002, *Pariah*, London, Verso.
70 See H. Young 1999, *This Blessed Plot: Britain and Europe from Churchill to Blair*, London, Macmillan.
71 J. K. Galbraith 1958, *The Affluent Society*, Harmondsworth, Penguin.
72 It was an ambiguous mix of elements – part parasitic upon the mainstream consumer capitalism (the mainstream was rich enough to tolerate drop-outs, who moreover generated their own economic niche products), part oppositional (for example, Herbert Marcuse on 'repressive tolerance' or the early 'ecology movement' and subsequently, after their success in Germany, 'the greens') and part simply media-generated fashion (hippies, etc.)).
73 R. Hilberg 2003 (3rd edn), *The Destruction of the European Jews*, Yale University Press; Finkelstein 2001.
74 See Judt 2002.
75 B. Laffan, R. O'Donnell and M. Smith 2000, *Europe's Experimental Union: Rethinking Integration,* London, Routledge; J. Richardson (ed.) 2001, *European Union: Power and Policy Making*, London, Routledge; B. Rosamund 2000, *Theories of European Integration*, London, Palgrave.
76 Laffan *et al.* 2000.
77 This is an ambiguous destination and either way, not a part of the politics of the public sphere. See J. G. Ballard 1996, *Cocaine Nights*, London, Flamingo; J. G. Ballard 2000, *Super-Cannes*, London, Flamingo; J. G. Ballard 2003, *Millennium People*, London, Flamingo; Zadie Smith 2000, *White Teeth*, Harmondsworth, Penguin.

11 Afterword: subaltern dreams

I have argued elsewhere that social theorizing must be reflexively embedded within the social process which it endeavours to grasp and understand,[1] in which case my own intellectual and social location might be noted, my professional identity. The great sociologist C. Wright Mills argued that it was the mix of private concerns and public issues which sparked the scholarly imagination,[2] opening up lines of research and reflection: it is here that the theorist begins to make a contribution. My own interests have multiple roots in childhood, travel and politics and they revolve around the ways in which people read and understand change. In childhood my thinking was shaped by the patterns of ideas current within my family and local community. There were patterns of parental deference, figures picked out as having authority – I wondered why some people rather than others were chosen. There were stories of war, there were aspirations (material and social) and these were presented to me as obvious and urgent goals. There are also memories from childhood that run wider, touching on the public sphere: a Union Jack hanging from a window in our house (Queen Elizabeth's coronation); an election poster glimpsed from the back of the family car on the way from shopping; Harold Macmillan's 'you've never had it so good'. Much of this is the intellectual and moral territory of the respectable working and lower middle classes in the 1950s and early 1960s. University days presented more ideas to explore, the territory of the arts and social sciences, with sharp debates about the Vietnam War and ideas of an environmentally sustainable society. These styles of reflection continued; they became my trade. It is a trade shaped by travel – I have lived and worked in East Asia and Germany and enjoyed all of it, learning day by day. The experience of other cultures, other lives, other ways of 'slicing up the world',[3] has been important. The experiences in themselves were unremarkable, but these were patterns of life outside the cultural reach of Britain/Britishness. Thus I had visited Germany as a child on school exchange visits, around 1960, pre-Beatles. At that time I had a head filled with images passed on from my parents' generation – centrally the experience of war – but I found an ordered place. In the window of the food hall of a departmental store in the small town where I stayed I saw a spread of marvellous cakes – the stereotypes and the cakes clashed, and I forgot the former and went with the latter. I have enjoyed Germany ever since. But it was in East Asia that I met a culture radically other than my own: Singapore, later Tokyo and Hong Kong. New people, food and climates, all dazzling and exciting cities. In these places I described myself as English and was accepted as a hard-working professional, an achieved not ascribed status. In other words I have lived outside the cultural bubble of Britain/Britishness, in particular, free of the British class system. It is to live without a weight on one's shoulders. I have also found the 'English abroad' to be as open-minded, energetic and optimistic as any other group – with some exceptions – but the positive impression has remained.

I happened to be in Germany in 1988/89. It was an interesting period. In 1988 there were elections to the European parliament. The German population took these elections seriously. In the same period a particular English view of Europe was offered at the European Cup Finals, by football hooligans. In the spring and summer of 1989 Presidents Bush (the first) and Gorbachev enjoyed triumphant visits and there was a growing stream of refugees from East Germany who travelled via Hungary to Germany, crossing the border at Passau to be greeted with great enthusiasm. In autumn 1989 the Berlin Wall was opened and there was a sequence of revolutions and reforms running through the countries of Central Europe, the wheel of history visibly turning, a dramatic and exciting time. On my return to Britain I discovered a curiously uninvolved response. The attitude was quite striking and I investigated. It picked up a long indirect interest in politics, in the patterns of power and ideology within the country, the ideas carried in institutions, patterns of ordinary life and published texts. The range of available patterns of understanding was familiar. The dominant position of the official ideology of Britishness was clear and it unpacked in multifarious ways through the lives of ordinary people. In contrast, Englishness seemed muted and muddled, a mix of nostalgia ('heritage') and the unremarked routines of ordinary life.

It seems to me – at present – that there are strong structural demands for change in the UK. The events and implications of 1989/91–2003 cannot be ignored. Current patterns of change will radically disturb the British polity: the elite will resist, but the masses have an opportunity. The questions are of recognition and action. The demands of circumstances can register with the agent in one of two ways – obligatory or voluntary: in the first case the agent reacts as soon as the situation is understood;[4] in the latter case we grant that there are many situations in which recognition can be distanced from the agent[5] – space can be made to grant the claims, to reject them or to subvert them.[6] Changing circumstances generate changing understandings that require novel lines of action – a mix of accommodation to the changes coupled with the use of that room for the choice that is available. Our current phase can be read in terms of contemporary concerns shaped by the residues, memories and legacies of earlier phases. The changes are difficult to grasp, difficult to order and difficult to live through. The upheavals of 1989/91–2003 mean that the location of Britain within the modern global system is now open to critical inspection in a way not available during periods of relative stability. The context within which the polity must operate comprises the internationalized global system, the pattern of regions and the locally dominant project of the European Union. Similarly, just as the location of the polity is open to inspection, so too are the goals affirmed. The post-Second World War elite preference for the status of an American subordinate is not merely not the only option on offer, it is, given the extant involvement with the EU, the more implausible one.[7] The changes running through the circumstances surrounding the polity point – in my view – towards involvement in Europe.

So, I return to the issue of the patterns of understanding current within my community. There are many available arguments for reform (indeed, there is a veritable industry making the arguments), and travel offers useful comparisons. However, I cannot see progressive change happening within the confines of the political-cultural project of Britain. The structure of the British state is oligarchic, power is centralized and ruling groups reproduce themselves as the establishment. The official ideology is centrally liberal, celebrating the particularity of the British and having effective informal extension in a spread of ideas that run through the social world. It is designed to demobilize and it is largely successful. It has produced an emotivist[8] political culture of subjective protest confronting pseudo-scientific bureaucratic expertise. It means that popular energy runs in other channels: family and friends; consumption; clubs and societies; and local communities. It is rich and vigorous. There are many pleasures. It falls short, however, of a rational democratic – adult – political life. It seems to me that change will only be impressed on the polity. It is an unsettling thought. Barrington Moore[9] has detailed the response of the Qing Dynasty to the demands of change. In the nineteenth century the expansionary dynamics of industrial capitalism brought European traders to China – they were insistent, troublesome and powerful. The Qing authorities responded with a mixture of incomprehension, inaction and a broad desire for the problems to go away. They did not, and rather the problems multiplied. By the early years of the twentieth century the Qing authorities began to respond but by that time it was too late, there was no constituency for reform and domestic Chinese opponents did not want a successful dynasty. What they wanted, and what they got, was the collapse of the regime in an optimistic forward-looking republican revolution. Events subsequently turned sour with domestic reaction, conflict and foreign intervention. The movement of China into the modern world was much delayed, at great cost to its people (and the various elites, both reactionary and progressive). Moore points out the lesson: ruling groups have to read and react to structural change; engagement is rational and denial is not an option.

So what, finally, are my subaltern dreams? They are of effective change. The clear route to the future for the population is Europe. But this is not an assemblage of piecemeal reforms, it is a complex package – 'a great collective project'[10] – it implies democratization, Europeanization and modernization. It is a vaulting aspiration: catching up and joining in the mainstream. The key is political reform. The EU has a core membership that is wedded to the idea of democracy. The democratization of Britain would entail reaffirming a project that was blocked in the early nineteenth century,[11] the shift from an essentially oligarchic liberal and deferential polity to a contemporary variant of republican democracy. Some of these changes are inevitable. Some involve choice. What is interesting about the project of Europe is that it offers a novel political and intellectual space. It escapes the grasp of the British state-regime and its official ideology. It is free of the dead weight of 'Britain' and 'Britishness'. It opens up the world beyond the bubble. In another text[12] I have argued that the nature of high modernity fosters cosmopolitanism, adulthood and scepticism, and I speculate that an English variant might be found within the wider context of the project of the European Union – this is the heart of my 'subaltern dreams', the business of 'relocating England'.

However, I can see no further into the future than any one else. The processes identified are dynamic and their outcome contingent. Contemporary change in Europe could serve to emancipate an optimistic, creative, amiable people, the English, but the British polity is resilient – it has resisted change before and is doing so now. So a final word; don't hold your breath!

Notes

1. P. W. Preston 1985, *New Trends in Development Theory*, London, Routledge.
2. C. W. Mills 1970, *The Sociological Imagination*, Harmondsworth, Penguin, annexe.
3. Jonathan Culler 1976, *Saussure*, London, Fontana. Culler uses this phrase to illuminate the active nature of language – it does not describe, it constitutes.
4. P. Winch 1958 (2nd edn 1990), *The Idea of a Social Science and its Relation to Philosophy*, London, Routledge. A new concept implies a new social world.
5. A. MacIntyre 1962, 'A Mistake about Causality in Social Science', in P. Laslett and W. G. Runciman (eds) *Philosophy, Politics and Society, Series 2*, Oxford, Blackwell.
6. J. S. Scott 1985, *Weapons of the Weak*, Yale University Press.
7. A particular debate, discussed by A. Gamble 2003, *Between Europe and America: The Future of British Politics*, London, Palgrave.

8 A. MacIntyre 1981, *After Virtue: A Study in Moral Theory*, London, Duckworth.

9 B. Moore 1966, *The Social Origins of Dictatorship and Democracy: Lord and Peasant in the Making of the Modern World*, London, Allen Lane.

10 F. Jameson 1991, *Postmodernism: Or The Cultural Logic of Late Capitalism*, London, Verso.

11 See D. Marquand 1988, *The Unprincipled Society*, London, Fontana, and T. Nairn 1988, *The Enchanted Glass*, London, Radius.

12 P. W. Preston 1997, *Political-Cultural Identity: Citizens and Nations in a Global Era*, London, Sage.

Bibliography

Ackroyd, P. 2000, *London: The Biography*, London, Chatto.
Addison, P. 1977, *The Road to 1945*, London, Jonathan Cape.
Addison, P. 1995 (2nd edn), *Now the War is Over: A Social History of Britain 1945–51*, London, Pimlico.
Albrow, M. 1970, *Bureaucracy*, London, Macmillan.
Ali, M. 2003, *Brick Lane*, London, Doubleday.
Alibai-Brown, Y. 2001, *Who Do We Think We Are: Imagining the New Britain*, Harmondsworth, Penguin.
Anderson, B. 1983, *Imagined Communities*, London, Verso.
Anderson, P. 1992, *English Questions*, London, Verso.
Anderson, P. 2002, 'Force and Consent', *New Left Review*, 17 Sept/October.
Anderson, P. 2003, 'Casuistries of Peace and War', *London Review of Books*, 6 March.
Aron, R. 1973, *The Imperial Republic: The US and the World 1945–1973*, London, Weidenfeld.
Ascherson, N. 1988, *Games with Shadows*, London, Radius.
Ashford, N. 1992, 'The Political Parties' in S. George (ed.), *Britain and the European Community: The Politics of Semi-Detachment*, Oxford University Press.
Auden, W. H. *September 1st 1939*.
Backhouse, R. 2002, *The Penguin History of Economics*, Harmondsworth, Penguin.
Ballard, J. G. 1988, *Empire of the Sun*, London, Grafton.
Ballard, J. G. 1996, *Cocaine Nights*, London, Flamingo.
Ballard, J. G. 2000, *Super-Cannes*, London, Flamingo.
Ballard, J. G. 2003, *Millennium People*, London, Flamingo.
Barnes, J. 1998, *England, England*, London, Picador.
Barraclough, G. 1964, *An Introduction to Contemporary History*, Harmondsworth, Penguin.
Bauman, Z. 1973, *Culture as Praxis*, London, Routledge.
Bauman, Z. 1988, *Freedom*, Open University Press.
Bauman, Z. 1988, *Legislators and Interpreters*, Cambridge, Polity.
Bauman, Z. 1989, *Modernity and the Holocaust*, Cambridge, Polity.
Berger, P. and Luckmann, T. 1966, *The Social Construction of Reality*, Harmondsworth, Penguin.

Berghahn, V. R. 1988, *Modern Germany: Society, Economy and Politics in the Twentieth Century*, Cambridge University Press.

Berlin, I. 1969, *Four Essays on Liberty*, Oxford University Press.

Billig, M. 1995, *Banal Nationalism*, London, Sage.

Bobbitt, P. 2002, *The Shield of Achilles: War, Peace and the Course of History*, New York, Alfred Knopf.

Braithwaite, R. 2003, *Prospect* March/April.

Brewer, A. 1980, *Marxist Theories of Imperialism*, London, Routledge.

Bridge, C. and Fedorowich, K. (eds) 2003, *The British World: Diaspora, Culture and Identity*, London, Frank Cass.

Bulmer, S. 1992, 'Britain and the European Community' in S. George (ed.), *Britain and the European Community: The Politics of Semi-Detachment*, Oxford University Press.

Buruma, I. 1999, *Voltaire's Coconuts, or Anglomania in Europe*, London, Weidenfeld.

Calvocoressi, P. 1997, *Fall Out: World War II and the Shaping of Postwar Europe*, London, Longman.

Campbell, B. 1984, *Wigan Pier Revisited*, London, Virago.

Canadine, D. 1983, 'The Context, Performance and Meaning of Ritual: the British Monarchy and "the Invention of Tradition"' in E. Hobsbawm and T. Ranger (eds), *The Invention of Tradition*, Cambridge, Canto.

Canadine, D. 1998, *Class in Britain*, Harmondsworth, Penguin.

Canadine, D. 2001, *Ornamentalism: How the British Saw their Empire*, London, Allen Lane.

Carr, E. H. [1939] 2001, *The Twenty Years Crisis*, London, Palgrave.

Castenada, C. 1971, *A Separate Reality*, London: The Bodley Head.

Cecchini, P. 1988, *The European Challenge 1992: The Benefits of a Single Market*, London, Wildwood.

Christiansen, T., Jorgensen, K. E. and Wiener, A. (eds) 2001, *The Social Construction of Europe*, London, Sage.

Chua B. H. 1995, *Communitarian Ideology and Democracy in Singapore*, London, Routledge.

Clairmonte, F. 1960, *Economic Liberalism and Underdevelopment*, Bombay, Asia Publishing House.

Clammer, J. 1995, *Difference and Modernity: Social Theory and Contemporary Japanese Society*, London, Kegan Paul International.

Clark, J. C. D. 2003, *Our Shadowed Present: Modernism, Postmodernism and History*, London, Atlantic Books.

Cohen, A. P. 1994, *Self-Consciousness*, London, Routledge.

Cohen, N. 2004, *Pretty Straight Guys*, London, Faber and Faber.

Cohen, S. 1972, *Folk Devils and Moral Panics*, Oxford, Martin Robertson.

Cole, K., Cameron, J. and Edwards, C. 1991, *Why Economists Disagree*, London, Longman.

Colley, L. 1992, *Britons: Forging the Nation 1707–1837*, Yale University Press.

Colls, R. 2002, *Identities of England*, Oxford University Press.

Cook, R. 2003, *The Point of Departure*, London, The Free Press.

Crawford, B. M. A. 2000, *Idealism and Realism in International Relations*, London, Routledge.

Crick, B. 1980, *George Orwell; A Life*, Harmondsworth, Penguin.

Croft, S. Redmond, J., Wyn Rees, G. and Webber, M. 1999, *The Enlargement of Europe*, Manchester University Press.

Culler, J. 1976, *Saussure*, London, Fontana.

Cummings, B. 1999, *Parallax Visions*, Duke University Press.

Darras, J. and Snowman, D. 1990, *Beyond the Tunnel of History*, London, Macmillan.

Davies, N. 1997, *Europe: A History*, London, Pimlico.

Davies, N. 2000, *The Isles: A History*, London, Papermac.

Davies, N. 2001, 'Britain and Australia: Holding Together or Falling Apart?' Public lecture at the *New South Wales Centenary Federation Committee Symposium, The Holding Together Programme.*

Delanty, G. 1995, *Inventing Europe: Ideas, Identity and Reality*, London, Macmillan.

Diez, T. 2001, 'Speaking Europe: The Politics of Integration Discourse' in T. Christiansen *et al.* (eds) *The Social Construction of Europe*, London, Sage.

Dower, J. 1999, *Embracing Defeat: Japan in the Aftermath of World War II*, London, Allen Lane.

The Economist.

Easthope, A. 1999, *Englishness and National Culture*, London, Routledge.

Eco, U. 1987, *Travels in Hyper-reality*, London, Picador.

Edwards, G. 1992, 'Central Government', in S. George (ed.) *Britain and the European Community: The Politics of Semi-Detachment*, Oxford University Press.

Ewing, K. D. and Gearty, C. A. 1990, *Freedom under Thatcher*, Oxford University Press.

Fay, B. 1987, *Critical Social Science*, Oxford, Blackwell.

Featherstone, M. 1991, *Consumer Capitalism and Postmodernism*, London, Sage.

Finkelstein, N. 2001, *The Holocaust Industry*, London, Verso.

Foucault, M. 1977, *Discipline and Punish: The Birth of the Prison*, New York, Pantheon.

Frank, A. G. 1983, *The European Challenge*, Nottingham, Spokesman Books.

Frank, A. G. 1998, *Re-Orient: Global Economy in the Asian Age*, University of California Press.

Freedman, L. 2003, *New Statesman* 24 March.

Fukuyama, F. 1992, *The End of History and the Last Man*, London, Hamish Hamilton.

Gadamer, H. G. 1960, *Truth and Method*, London, Sheed and Ward.

Gaddis, J. L. 1997, *We Now Know: Rethinking Cold War History*, Oxford University Press.

Galbraith, J. K. 1958, *The Affluent Society*, Harmondsworth, Penguin.

Galbraith, J. K. 1975, *The Great Crash*, Harmondsworth, Penguin.

Gamble, A. 1988, *The Free Economy and the Strong State*, London, Macmillan.

Gamble, A. 2003, *Between Europe and America: The Future of British Politics*, London, Palgrave.

Garton-Ash, T. 1990, *We the People: The Revolution of 89*, London, Granta.

Gellner, E. 1964, *Thought and Change*, London, Wiedenfeld.

Gellner, E. 1983, *Nations and Nationalism*, Oxford, Blackwell.

Gellner, E. 1991, *Plough, Sword and Book*, London, Paladin.

Gellner, E. 1992, *Reason and Culture*, Oxford, Blackwell.

George, S. (ed.) 1992, *Britain and the European Community: The Politics of Semi-Detachment*, Oxford University Press.

Giddens, A. 1976, *New Rules of Sociological Method*, London, Hutchinson.

Giddens, A. 1979, *Central Problems in Social Theory*, London, Macmillan.
Giddens, A. 1984, *The Constitution of Society*, Cambridge, Polity.
Giddens, A. 1991, *Modernity and Self-Identity*, Cambridge, Polity.
Giddens, A. 2002, *Runaway World: How Globalization is Shaping our Lives*, London, Profile.
Gilson, J. 2000, *Japan and the European Union*, London, Macmillan.
Ginsborg, P. 1990, *A History of Contemporary Italy*, Harmondsworth, Penguin.
Glenny, M. 1992, *The Fall of Yugoslavia: The Third Balkan War*, Harmondsworth, Penguin.
Gramsci, A. 1971, *Selections from the Prison Notebooks*, edited and translated by Q. Hoare and G. N. Smith, London, Lawrence and Wishart.
Grass, G. 2000, *My Century*, London, Faber and Faber.
Grass, G. 2002, *Crabwalk*, London, Faber and Faber.
Gray, J. 2003, *New Statesman* 21 April.
Griffiths, R. 1983, *Fellow Travellers of the Right*, Oxford University Press.
Guardian.
Gudeman, S. 1986, *Economics as Culture*, London, Routledge.
Haas, E. 1958, *The Uniting of Europe*, Stanford University Press.
Haas, E. 1964, *Beyond the Nationstate: Functionalism and International Organisations*, Stanford University Press.
Habermas, J. 1992, 'Citizenship and National Identity: Some Reflections on the Future of Europe', *Praxis International* 12.
Hall, S. 1990, *The Hard Road to Renewal: Thatcherism and the Crisis of the Left*, London, Verso.
Halliday, F. 1989, *Cold War, Third World*, London, Radius.
Harvie, C. 1992, *Cultural Weapons: Scotland and Survival in the New Europe*, Edinburgh, Polygon.
Haslam, J. 1999, *The Vices of Integrity: E. H. Carr 1892–1982*, London, Verso.
Hawthorn, G. 1976, *Enlightenment and Despair*, Cambridge University Press.
Hay, C. 1996, *Restating Social and Political Change*, Open University Press.
Hay, C. 1997, 'Blaijorism: Towards a One Vision Polity?', *Political Quarterly* 68, 372–9.
Hay, C. 1999, *The Political Economy of New Labour*, Manchester University Press.
Hay, C. 2002, *Political Analysis*, London, Palgrave.
Hayek, F. 1944, *The Road to Serfdom*, London, Routledge.
Held, D. 1987, *Models of Democracy*, Cambridge, Polity.
Held, D. and McGrew, A. 2002, *Globalization/Anti-Globalization*, Cambridge, Polity.
Heller, J. 1964, *Catch 22*, London, Corgi.
Hellman, L. 1976, *Scoundrel Time*, New York, Little, Brown.
Hennessey, P. 1989, *Whitehall*, New York, Free Press.
Hennessy, P. 1992, *Never Again: Britain 1945–1951*, London, Jonathan Cape.
Higgot, R. and Robinson, R. (eds) 1985, *Southeast Asia: Essays in the Political Economy of Structural Change*, London, Routledge.
Hilberg, R. 2003 (3rd edn), *The Destruction of the European Jews*, Yale University Press.
Hirst, P. 1990, *New Statesman and Society* 16 November.
Hirst, P. and Thompson, G. 1999 (2nd edn), *Globalization in Question*, Cambridge, Polity.
Hitchcock, W. I. 2003, *Prospect* April.
Hitchens, C. 1990, *Blood, Class and Nostalgia: Anglo-American Ironies*, London, Vintage.

Hitchens, P. 2000, *The Abolition of Britain*, London, Quartet.
Hobsbawm, E. 1994, *The Age of Extremes: The Short Twentieth Century 1914–1991*, London, Michael Joseph.
Hobsbawm, E. and Ranger, T. (eds) 1983, *The Invention of Tradition*, Cambridge, Canto.
Hoffman, S. 1966, 'Obstinate of Obsolete: The State in Western Europe', *Deadalus*, 95.
Hoggart, R. 1958, *The Uses of Literacy*, Harmondsworth, Penguin.
Hoggart, R. 1989, *A Local Habitation (Life and Times Vol 1: 1918–1940)*, Oxford University Press.
Hoggart, R. 1995, *The Way We Live Now*, London, Chatto.
Holland, M. 2002, *The European Union and the Third World*, London, Palgrave.
Honderich, T. 1990, *Conservatism*, London, Hamish Hamilton.
Hoopes, T. 1997, *FDR and the Creation of the UN*, Yale University Press.
Hughes, R 1991, *The Shock of the New*, London, Thames and Hudson.
Hughes, S. 1979, *Consciousness and Society: The Reorientation of European Social Thought 1890–1930*, Brighton, Harvester.
Huntington, S. P. 1993, 'The Clash of Civilizations', *Foreign Affairs* 72/3.
Hutton, W. 1996, *The State We're In*, London, Vintage.
Hutton, W. 2002, *The World We're In*, London, Little, Brown.
Ignatieff, M. 1994, *Blood and Belonging: Journeys into the New Nationalism*, London, Vintage.
Inglis, F. 1993, *Cultural Studies*, Oxford, Blackwell.
Ishiguro, K. 1987, *An Artist of the Floating World*, London, Faber and Faber.
Ishiguro, K. 1989, *The Remains of the Day*, London, Faber and Faber.
James, C. L. R. 1973, *World Revolution*, Westport CT, Hyperion Press.
James, C. L. R. 1984, *Beyond a Boundary*, New York, Pantheon.
James, H. 1994, *A German Identity: 1770 to the Present Day*, London, Phoenix.
Jameson, F. 1991, *Postmodernism: Or the Cultural Logic of Late Capitalism*, London, Verso.
Jay, M. 1973, *The Dialectical Imagination*, Boston MA, Little, Brown.
Jenks, C. 1983, *Culture*, London, Routledge.
Jennings, P. 1986, *The Living Village*, London, Hodder.
Jessop, B., Bonnet, K., Bromley, S. and Ling, T. 1988, *Thatcherism: A Tale of Two Nations*, Cambridge, Polity.
Johnson, C. 1992, *MITI and the Japanese Miracle: The Growth of Industrial Policy, 1925–1975*, Tokyo, Tutle.
Johnson, C. 2000, *Blowback: The Costs and Consequences of American Empire*, New York, Little Brown.
Johnson, C. 2004, *The Sorrows of Empire: Militarism, Secrecy and the End of the Republic*, London, Verso.
Jones, C. 1998, *E. H. Carr and International Relations: A Duty to Lie*, Cambridge University Press.
Jones, E. 2003, *The English Nation: The Great Myth*, Stroud, Sutton Publishing.
Judt, J. 2002, 'The Past is Another Country: Myth and Memory in Post-war Europe' in J. W. Muller (ed.) *Memory and Power in Post War Europe*, Cambridge University Press.
Kampfer, J. 2003, *Blair's Wars*, London, The Free Press.
Kaye, H. 1984, *The British Marxist Historians*, Cambridge, Polity.
Keane, J. 1988, *Democracy and Civil Society*, London, Verso.

Kee, R. 1984, *1939 The World We Left Behind*, London, Weidenfeld.

Keohane, R. and Nye, J. 1977, *Power and Interdependence, World Politics in Transition*, Boston, Little, Brown.

Kerr, P. 2001, *Postwar British Politics*, London, Routledge.

Kolko, G. 1968, *The Politics of War*, New York, Vintage.

Kornhauser, A. 1958, *The Politics of Mass Society*, London, Collier.

Kumar, K. 2003, *The Making of English National Identity*, Cambridge University Press.

Kureishi, H. 1990, *The Buddha of Suburbia*, London, Faber and Faber.

Laffan, B., O'Donnell, R. and Smith, M. 2000, *Europe's Experimental Union: Rethinking Integration*, London, Routledge.

Laidi, Z. 1998, *A World Without Meaning in International Politics*, London, Routledge.

Langford, P. 2000, *Englishness Identified: Manners and Character 1650–1850*, Oxford University Press.

Laslett, P. and Runciman, W. G. (eds) 1962, *Philosophy, Politics and Society, Series 2*, Oxford, Blackwell.

LeCarre, J. 1960, *A Small Town in Germany*, London, Pan.

Lee, L. 1962, *Cider with Rosie*, Harmondsworth, Penguin.

Lindquist, S. 2002, *A History of Bombing*, London, Granta.

Lodge, J. (ed.) 1989, *The European Community: The Challenge of the Future*, London, Pinter.

Long, N. (ed.) 1992, *Battlefields of Knowledge*, London, Routledge.

Longford, P. 2000, *Englishness Identified: Manners and Character 1650–1850*, Oxford University Press.

Lucas, S. 2003, *Orwell*, London, Haus Publishing.

Lukes, S. 1973, *Individualism*, New York, Harper and Row.

MacAlister, R. 1997, *From EC to EU: An Historical and Political Survey*, London, Routledge.

MacIntyre, A. 1962, 'A Mistake about Causality in the Social Sciences' in P. Laslett and Runciman, W. G. (eds) *Philosophy, Politics and Society, Series 2*, Oxford, Blackwell.

MacIntyre, A. 1985, *After Virtue: A Study in Moral Theory*, London, Duckworth.

Macmillan, M. 2003, *Peacemakers: Six Months that Changed the World*, London, John Murray.

Macpherson, C. B. 1962, *The Political Theory of Possessive Individualism*, Oxford University Press.

Macpherson, C. B. 1973, *Democratic Theory: Essays in Retrieval*, Oxford University Press.

Mann, M. 2003, *Incoherent Empire*, London, Verso.

Mantel, H. 2003, *Giving Up the Ghost: A Memoir*, London, Fourth Estate.

Marcussen, M. *et al.* 2001, 'Constructing Europe: The Evolution of Nation-State Identities', in T. Christiansen *et al.* (eds) *The Social Construction of Europe*, London, Sage.

Marquand, D. 1988, *The Unprincipled Society*, London, Fontana.

Marquand, D. 2003, *New Statesman* March.

Marquand, D. 2003, *New Statesman* 24 November.

Marquand, D. 2004, *Decline of the Public*, Cambridge, Polity.

Marquand, D. 2004, *New Statesman* 19 January.

Marquese, M. 1994, *Anyone but England: Cricket and the National Malaise*, London, Verso.

Marquese, M. 2003, *Chimes of Freedom: The Politics of Bob Dylan's Art*, London, The New Press.
Mayer, A. 1981, *The Persistence of the Old Regime*, New York, Croom Helm.
Mayer, A. 1988, *Why Did the Heavens Not Darken: The 'Final Solution' in History*, New York, Pantheon.
Mazower, M. 1998, *Dark Continent: Europe's Twentieth Century*, New York, Alfred Knopf.
Mearsheimer, J. J. and Walt, S. 2003, 'An Unnecessary War', *Prospect* March.
Merquior, J. G. 1985, *Foucault*, London, Fontana.
Middlemas, K. 1979, *The Politics of Industrial Society*, London, Andre Deutsche.
Miliband, R. 1982, *Capitalist Democracy in Britain*, Oxford University Press.
Miliband, R. 1972 (2nd edn), *Parliamentary Socialism*, London, Merlin.
Miliband, R. 1973, *The State in Capitalist Society*, London, Quartet.
Mills, C. W. 1970, *The Sociological Imagination*, Harmondsworth, Penguin.
Mitrany, D. 1966, *A Working Peace System*, Chicago, Quadrangle.
Mitrany, D. 1975, *The Functional Theory of Politics*, London, Martin Robertson.
Moore, B. 1966, *The Social Origins of Dictatorship and Democracy: Lord and Peasant in the Making of the Modern World*, London, Allen Lane.
Moravcsik, A. 1998, *The Choice for Europe*, London, UCL Press.
Nairn, T. 1977, *The Break-up of Britain*, London, New Left Books.
Nairn, T. 1988, *The Enchanted Glass*, London, Hutchinson Radius.
Nairn, T. 2002, *Pariah*, London, Verso.
New Statesman.
Newman, M. 2002, *Ralph Milliband and the Politics of the New Left*, London, Merlin.
Orwell, G. 1941, *The Lion and the Unicorn*, Harmondsworth, Penguin.
Overbeek, H. 1990, *Global Capitalism and National Decline*, London, Allen and Unwin.
Palmer, J. 1988, *Europe without America*, Oxford University Press.
Parkin, F. 1972, *Class Inequality and Political Order*, London, Paladin.
Paxman, J. 1990, *Friends in High Places*, Harmondsworth, Penguin.
Paxman, J. 1999, *The English: A Portrait of a People*, Harmondsworth, Penguin.
Pienning, C. 1997, *Global Europe: The European Union in World Affairs*, Boulder, Lynne Reinner.
Pierson, C. 2001, *Hard Choices: Social Democracy in the 21st Century*, Cambridge, Polity.
Pinder, J. 1989, 'The Single Market: A Step Towards a European Union' in J. Lodge (ed.) *The European Community: The Challenge of the Future*, London, Pinter.
Pinder, J. 1991, *The European Community: The Building of a Union*, Oxford University Press.
Popper, K. 1957, *The Poverty of Historicism*, London, Routledge.
Porter, P. 1992, *Myths of the English*, Cambridge, Polity.
Porter, R. 2000, *Enlightenment: Britain and the Creation of the Modern World*, London, Allen Lane.
Prebisch, R. 1950, *The Economic Development of Latin America and its Principle Problems*, New York, United Nations.
Preston P. 2003, *Guardian* March/April.
Preston, P. W. 1981, *Theories of Development*, London, Routledge.
Preston, P. W. 1985, *New Trends in Development Theory*, London, Routledge.

Preston, P. W. 1994, *Discourses of Development: State, Market and Polity in the Analysis of Complex Change*, Aldershot, Avebury.

Preston, P. W. 1994, *Europe, Democracy and the Dissolution of Britain*, Aldershot, Dartmouth.

Preston P. W. 1996, *Development Theory*, Oxford, Blackwell.

Preston, P. W. 1997, *Political-Cultural Identity: Citizens and Nations in a Global Era*, London, Sage.

Preston, P. W. 1998, *Pacific Asia in the Global System*, Oxford, Blackwell.

Preston, P. W. 2002, '9/11: Making Enemies; Some Uncomfortable Lessons for Europe'. Paper presented to the *European Union in International Affairs Conference*, Australian National University, 3/4 July.

Preston, P. W. and Gilson, J. (eds) 2001, *The European Union and East Asia*, Cheltenham, Edward Elgar.

Reuschemeyer, D., Huber-Stevens, E. and Stevens, J. D. 1991, *Capitalist Development and Democracy*, Cambridge, Polity.

Richardson, J. (ed.) 2001, *European Union: Power and Policy Making*, London, Routledge.

Rieff, D. 1995, *Slaughterhouse: Bosnia and the Failure of the West*, London, Vintage.

Robbins, D. 1991, *The Work of Pierre Bourdieu*, Open University Press.

Robbins, K. 1998, *Great Britain: Identities, Institutions and the Idea of Britishness*, London, Longman.

Rosamund, B. 2000, *Theories of European Integration*, London, Palgrave.

Rosamund, B. 2001, 'Discourses of Globalization and European Identities' in T. Christiansen *et al.* (eds) *The Social Construction of Europe*, London, Sage.

Rude, G. 1981, *The Crowd in History*, London, Wiley.

Ryder, J. and Silver, H. 1985, *Modern English Society*, London, Methuen.

Sage, L. 2001, *Bad Blood*, London, Fourth Estate.

Sampson, A. 1992, *The Essential Anatomy of Britain*, London, Hodder and Stoughton.

Sampson, A. 2004, *Who Runs this Place? The Anatomy of Britain in the 21st Century*, London, John Murray.

Samuel, R. 1994, *Theatres of Memory Vol 1: Past and Present in Contemporary Culture*, London, Verso.

Samuel, R. 1999, *Island Stories: Unravelling Britain (Theatres of Memory, Volume II)*, London, Verso.

Saville, J. 1988, *The Labour Movement in Britain*, London, Faber and Faber.

Schama, S. 2003, *A History of Britain III: The Fate of Empire 1776–2001*, London, BBC.

Scott, J. C. 1977, 'Protest and Profanation: Agrarian Revolt and the Little Tradition', *Theory and Society* 4.1/4.2.

Scott, J. C. 1985, *Weapons of the Weak*, Yale University Press.

Scruton, R. 2001, *England: An Elegy*, London, Pimlico.

Sebald, W. G. 2004, *On the Natural History of Destruction*, Harmondsworth, Penguin.

Shaw, J. 'Postnational Constitutionalism in the European Union' in T. Christiansen *et al.* (eds) *The Social Construction of Europe*, London, Sage.

Shirer, W. L. 1960, *The Rise and Fall of the Third Reich*, London, Secker and Warburg.

Shore, C. 2000, *Building Europe: The Cultural Politics of European Integration*, London, Routledge.

Siedentop, L. 2001, *Democracy in Europe*, London, Allen Lane.
Simms, B. 2002, *Unfinest Hour: Britain and the Destruction of Bosnia*, Harmondsworth, Penguin.
Skidelsky, R. 2002, *John Maynard Keynes, A Biography: Fighting for Freedom* (Vol 3) London, Penguin.
Smith, Z. 2001, *White Teeth*, Harmondsworth, Penguin.
The Spectator.
Stiglitz, J. 2002, *Globalization and its Discontents*, London, Allen Lane.
Strange, S. 1988, *State and Markets*, London, Pinter.
Strange, S. 1989, *Casino Capitalism*, Oxford, Blackwell.
Thompson, E. P. 1968, *The Making of the English Working Class*, Harmondsworth, Penguin.
Thorne, C. 1986, *The Far Eastern War: States and Societies 1941–45*, London, Counterpoint.
Todd, E. 2003, *After the Empire: The Breakdown of the American Empire*, Columbia University Press.
Townsend, P. 1979, *Poverty in the United Kingdom*, University of California Press.
Traynor, I. 2003, *Guardian*, 28 April.
Urwin, D. 1997, *A Political History of Western Europe Since 1945*, London, Longman.
van der Pijl, K. 1984, *The Making of an Atlantic Ruling Class*, London, Verso.
Wade, R. 1990, *Governing the Market*, Princeton University Press.
Wallace, W. 1990, *New Statesman and Society* 9 November.
Wallace, W. 1990, *The Transformation of Western Europe*, London, Pinter.
Warwick, D. 1974, *Bureaucracy*, London, Longman.
Waterhouse, K. 1994, *City Lights: A Street Life*, London, Sceptre.
Waugh, E. 1962, *Brideshead Revisted*, Harmondsworth, Penguin.
Weight, R. 2002, *Patriots: National Identity in Britain 1940–2000*, London, Pan.
Weiss, L. 1999, *The Myth of the Powerless State: Governing the Economy in a Global Era*, Cambridge, Polity.
Wessels, W., Maurer, A. and Mittag, J. (eds) 2003, *Fifteen into One: The European Union and its Member States*, Manchester University Press.
Whitebrook, M. 2001, *Identity, Narrative and Politics*, London, Routledge.
Willliams, B. 1972, *Morality*, Harmondsworth, Penguin.
Williams, G. 1985, *When Was Wales?*, Harmondsworth, Penguin.
Williams, R. 1961, *The Long Revolution*, London, Chatto.
Williams, R. 1980, *Keywords*, London, Fontana.
Williams, R. 1958, *Culture and Society*, Harmondsworth, Penguin.
Winch, P. 1958 (2nd edn 1990), *The Idea of a Social Science and its Relation to Philosophy*, London, Routledge.
Wolf, J. 1981, *The Social Production of Art*, London, Macmillan.
Worsley, P. 1984, *The Three Worlds: Culture and World Development*, London, Weidenfeld.
Wright, P. 1985, *On Living in an Old Country*, London, Verso.
Wright, P. 1993, *A Journey Through the Ruins*, London, Flamingo.
Wright, P. 1995, *The Village that Died for England: The Strange Story of Tyneham*, London, Jonathan Cape.

Young, H. 1999, *This Blessed Plot: Britain and Europe from Churchill to Blair*, London, Macmillan.

Young, M. 1958, *The Rise of the Meritocracy*, Harmondsworth, Penguin.

Index

Act/Treaty of Union (1707) 41, 53, 168
Adenauer, K. 132, 135, 192, 194
Africa 54, 130, 136, 144, 170
Alfred, King of Wessex 3, 40
Algeria 44
Ali, M. 17, 165
America 35, 51
American Revolution 41, 53
Amis, K. 17
Amsterdam Treaty 197
Anderson, B. 31, 49, 54, 164
Anderson, P. 54, 126
Angola 44
Aron, R. 188
Ascherson, N. 69
ASEAN 144
ASEM 144
Asia 108
Atlee, C. 58
Austria 57, 136, 147, 183, 196, 197
Automobile Association 19
Aznar, J. M. 147

Baghot, W. 84
Ballard, J. G. 165
Barstow, S. 18
Bauman, Z. 12, 28, 66, 97
Belgium 44, 143, 147
Benelux 132–3, 142
Bentham, J. 84
Berlin, I. 92
Berlin Wall 107, 196
Berlusconi, S. 147
Beveridge, W. 57, 64, 170
Bevin, E. 44, 58
Birmingham 90
Blaijorism 69, 72
Blair, Tony 35, 63–5, 109–11, 147, 172
Bolsheviks 57
Bonn 192
Bourdieu, P. 98
Bragg, M. 21
Braine, J. 18
Brandt, W. 135
Bretton Woods 58, 59, 62–3, 131, 136, 139, 188, 194
Brezhnev, L. 196
Brussels 133, 138–9, 190, 199
Bulgaria 107
Bush, G. 136, 205
Bush, G. W. R. 45, 110, 146, 188, 197

Cadbury 161
Cambell, B. 19
Canada 142
Canadine, D. 164
Canterbury 90
CAP 118
Carr, E. H. 57

CENTO 194
CFSP 141, 143, 145, 147
Chamberlain, N. 57
Charter 77 107
Chartists 42, 53, 54, 55, 169
Cheltenham 90
Chesshyre, R. 165
China 144, 206–7
Christendom 3
Churchill, W. S. 68
Civil War (Chinese) 56
Civil War (English) 3, 41, 83
Civil War (European) 42
Civil War (Russian) 57
Colley, L. 4, 32–3, 53
Commonwealth 4, 58, 106, 190, 193
Congo 44
Constable, J. 92
Cornwall 90
Cotonou Agreement 145
Cumbria 90
Cyprus 197
Czech Republic 136, 197
Czechoslovakia 107, 196

Davies, N. 3, 39, 40–1, 50–1, 122, 180, 189
DeGaulle, C. 132–5, 187, 192, 194–5
Delanty, G. 150
Delors, J. 136, 196
Deng Xiaoping 144
Denmark 135, 142
D'Estaing, G. 135, 137
Diggers 41, 52
Dorset 17
Drake, F. 93
Dresden 184
Dubczek, A. 195
Dunn, N. 162
Dylan, B. 163

East Asia 5, 42, 45, 54, 58, 119, 124, 130, 143, 144, 183, 196, 204–5
East Germany 205
East London 18
Eastern Europe 107
ECB 136
ECSC 132–3, 191
EDC 133, 142
EFTA 106, 118
EMU 135
Engels, F. 15
English Revolution 41, 52, 84
Erhard, L. 135, 192
ERM 110, 119
Estonia 197
Euratom 132, 191
Europe 35, 45, 58–9, 106, 108, 119, 124
European Economic Community 118, 132, 193
European Union 1, 5, 34, 45, 50, 63, 71–3, 109–11, 116–18, 121, 123, 130–9, 140–2, 145, 148–9, 150, 171–3, 179, 188, 191, 197–9
Ewing, K. D. 87

Featherstone, M. 98
Finland 136, 197
Foucault, M. 150
Fowles, J. 17
France 40, 51–2, 96, 109, 132–3, 137, 142–3, 147, 181–2, 187, 195
Frank, A. G. 143
Frankfurt School 98–9
French Revolution 41–2, 53
Friedman, M. 192

Galbraith, J. K. 163
Gamble, A. 70
GATT 58, 131
Gearty, C. A. 87
Geldorf, B. 163
General Strike 57
German Democratic Republic 107
GLC 161
Glorious Revolution 168
Gorbachev, M. 107, 136, 141, 195–6, 205
Gramsci, A. 73

Great Patriotic War 187
Great War 4, 42, 56–7, 71, 131, 158, 164, 180–2, 185, 187, 190–2, 195, 204–5
Greece 136
Guernica 184
Gulf War 197

Haas, E. 134
Habermas, J. 148
Hall, S, 66, 68, 70
Hapsburg 4, 42, 56
Hardy, T. 21
Harvie, C. 69
Hayek, F. 192
Heath, Edward 118
Hennessy, P. 66
Hill, C. 52
Hilton, R. 52
Hitchens, C. 67
Hobbes, T. 41
Hobsbawm, E. 52, 182
Hoffman, S. 135
Hoggart, R. 14, 16, 21, 23, 25, 27, 31, 66, 148
Hohenzollern 4, 42, 56
Holmes, S. 93
Home, A. D. 67
Honeker, E. 196
Hong Kong 205
Hood, R. 93
Howard, E. 161
Hughes, R. 107
Hundred Years War 3, 40, 50–1
Hungary 57, 107, 136, 196–7, 205
Hutton, W. 63

IBRD 58, 131, 189
ICC 45
Iceland 142
IMF 58, 131, 146, 189
India 44
Indo-China 44
Indonesia 44
Iraq War 5, 35, 45, 59, 69, 106–7, 110, 120, 147, 198
Ireland 41, 57, 135
Irish Free State 39
Ishiguro, K. 162
Italy 57, 132–3, 137, 142, 195

Jameson, F. 28, 94
Japan 42–3, 144, 197
Jenkins, R. 135, 196
Judt, T. 148, 185, 197

Kee, R. 180
Keisinger, K. G. 135
Keynes, J. M. 57, 62, 68
Kohl, H. 136
Kolko, G. 131
Krushchev, N. 189, 195
Kureishi, H. 17

Laffan, B. 139, 198
Latin America 54, 108, 130, 136, 144, 180
Lawrence, D. H. 17, 20–1, 67
League of Nations 43, 110, 131
Leavis, F. R. 164
LeCarre, J. 20–1, 162
Lee, L. 14
Leeds 21, 90
Letchworth 161
Levellers 41, 52, 169
Lever Bros. 161
Liverpool 90
Locke, J. 41
Lome Convention 144
London 17, 55, 60, 90, 123, 161
Luckacs, G. 98
Lutyens, E. 161
Luxembourg 143

Maastricht Treaty 45, 136–8, 143, 147, 192, 197
MacIntyre, A. 93–4, 98
MacMillan, H. 67, 118, 184, 204
Major, John 35, 45, 63, 69, 109, 118

Malta 136
Manchester 90
Mannheim, K. 57
Mantel, H. 14, 16
Marquand, D. 60, 84–5, 123, 126
Marshall Plan 44, 58, 189, 191
Marx, K. 54
Mayer, A. 56
Middle Ages 50
Middle East 54, 144, 146–7, 183
Middlemas, K. 85
Milliband, R. 87
Mills, C. W. 204
Mitrany, D. 134, 138
Mitterand, F. 132, 136
Monnet, J. 191–2
Moore, B. 149, 206
Morris, W. 161
Moscow 44, 56
Mozambique 44
Murdoch, I. 18

NAFTA 172
Naipaul, V. S. 165
Nairn, T. 28, 35, 41, 52, 54, 83, 123, 169
NATO 4, 58–9, 108, 111, 120, 136, 142–3, 172, 188, 194
New Deal 43, 182
New York 189
Newcastle 90
NHS 65
Nice Treaty 136–7, 197
Norman Conquest 40
North Africa 183
North America 4, 41, 45, 119, 124
Northumberland 90
Norway 135, 142
Norwich 90
Nottinghamshire 17
Nuremberg 186

Offa, King of Mercia 3, 40
Orwell, G. 19, 160
Osborne, J. 162
Ottoman 43
Overbeek, H. 119

Pacific War 56, 143
Paine, T. 169
Paris 58
Parkin, F. 33, 69, 97
Passau 205
People's Republic of China 144
Poland 107, 136–7, 197
Pompidou, G. 132, 135
Popper, K. 192
Porter, R. 53, 168
Portugal 44, 57, 136
Potsdam 187
Potter, D. 165
Prague Spring 195–6

Qing Dynasty 206

Raban, J. 165
Redfield, R. 82
Reformation 40–1, 51
Rhineland 123
Romanov 4, 42, 56
Roosevelt, F. D. 43, 58, 110, 132, 192
Rumsfeld, D. 172
Rushdie, S. 165
Russia 56, 190

Sage, L. 16, 20
Salt, T. 161
Samuel, R. 24
Schuman, R. 133, 191
Scotland 41–2, 123
Scott, J. C. 90
Scruton, R. 166–8
SEATO 194
Second Reform Act 87
Second World War 1, 43–6, 56–9, 66, 70, 83, 91, 106, 108, 117–9, 125, 130, 136, 144–5, 157–8, 162–4, 170, 180, 183, 187, 189, 194, 206
Shakespeare 3, 41
Sillitoe, A. 20, 162

Singapore 205
Single European Act 119, 136–8, 192, 196
Sino-Japanese War 56
Slovenia 136, 197
Smith, Z. 18
Solidarity 107
South Asia 144, 170
South Wales 17
Southeast Asia 144
Soviet Union 131, 148, 196
Spain 40, 57, 136–7
Spinetti, A. 136
Stalin, J. 189, 195
Stuarts 40, 41
Suez 4, 44, 59, 67, 118
Sweden 136, 197

Thatcher, Margaret 45, 63, 68, 109, 118–19, 163
Theroux, P. 165
Third World 61, 108, 145
Thompson, E. P. 52, 160, 169
Tokyo 205
Tower Hamlets 17
Treaty of Paris 133
Tressel, R. 20
Tudors 3, 40, 51–2

United Nations 45, 58, 59, 110, 147, 189
USA 4, 5, 43, 46, 58, 67, 93, 109, 111, 130, 137, 141–6, 172, 182, 186–7, 189, 190–5, 197–8
USSR 43, 57, 93, 107–8, 142, 182, 185–7, 190–1, 193, 196

Verdun 181
Versailles Treaty 43, 131
Vietnam 59, 135, 146, 194–5, 204

Wales 42, 123
Wall Street Crash 43
War of Independence (American) 32, 42
Washington 43–5, 56, 59, 121, 131, 189
Waterhouse, K. 19
Waugh, E. 15, 20, 162
Weber, M. 87
Weimar 131
Welwyn Garden City 161
West Indies 170
Westminster model 55, 59, 84–5
Whitehall/Westminster 60, 84, 109
Williams, G. 69
Williams, R. 21, 160, 162
Wilson, H. 68
Wilson, W. 110
Winch, P. 34, 73
Winchester 90
Wright, P. 18, 24, 30, 32, 89, 95–6, 161
WTO 109, 145–6

Yalta 187
Yeltsin, B. 107, 195–6